# Fractional Random Vibrations II

This two-volume set provides a comprehensive study of fractional random vibration from the perspective of theory and practice. Volume II provides the analytical expressions for the responses of seven classes of fractional vibration systems excited by six types of random processes.

It examines the responses of seven classes of fractional vibrations driven by five fractional processes, namely, fractional Gaussian noise, generalized fractional Gaussian noise, fractional Brownian motion, fractional Ornstein-Uhlenbeck process, the process with the von Kármán spectrum, and the process with the Pierson-Moskowitz spectrum. The present results apply the theories discussed in Volume I to the novel and cutting-edge topic of fractional processes, with a special focus on fractional vibration systems for marine structures influenced by the Pierson-Moskowitz and von Kármán spectra.

The title will be essential reading for students, mathematicians, physicists, and engineers interested in fractional random vibration phenomena.

**Ming Li** is a professor at Ocean College, Zhejiang University, China, and an emeritus professor at East China Normal University. He has been a contributor for many years to the fields of mathematics, statistics, mechanics, and computer science. His publications with CRC Press also include *Multi-Fractal Traffic and Anomaly Detection in Computer Communications*, *Fractal Teletraffic Modeling and Delay Bounds in Computer Communications*, and *Fractional Vibrations with Applications to Euler-Bernoulli Beams*.

## Fractional Order Thinking in Exploring the Frontiers of STEM

Fractional Order Thinking (FOT) is about solving today's complex problems in the physical, social and life sciences using the tools of fractional order calculus (FOC). Very soon after Mandelbrot introduced the fractal paradigm into the scientific lexicon it was shown that the integer order calculus (IOC) could not describe the dynamics of fractal processes. A new kind of calculus was required to construct the equations of motion for fractal dynamic processes which turned out to be fractional order Hamiltonian equations (FOHEs). The FOHEs are one tool in the FOC toolbox which concerns how to apply the operators of differentiation and integration of non-integer orders. Rejecting the fractional calculus is equivalent to saying there are no numbers between neighboring integers.

In this book series, we explore the core motivation of fractional calculus by first showing the "core motivation" of IOC invented by Newton and Leibnitz, the fundamental ideas of which can be traced back to the time of Heraclitus of Ephesus. Our ultimate message is that the "IOC" is driven by "the desire and the need" for the "quantification of changes" based on energy gradients in complex dynamic networks, whereas "FOC" is driven by "the desire and the need of understanding complexity" based on information gradients in those same networks. In science, technology, engineering, and mathematics (STEM), metaphorically speaking, there is plenty of room between the integers to enable the archiving of better than the best modelling, control performance, robustness, resilience, and even intelligence.

There is an increasing interest in fractional order dynamic systems (FODS) and controls in the recent research literature, not only because of their novelty but also due to their practical applications. But to accomplish all of this requires a new way of thinking, the Fractional Order Thinking (FOT) referred to earlier, which in turn must be preceded by a new STEM curriculum. This series will offer a unique platform to demonstrate such additional benefits in our improved understanding of complexity and stochastic dynamics via FOT.

This exciting book series features:

- A forum demonstrating the good consequences of using fractional order thinking (FOT).
- Fractional calculus concepts are presented in the context of complex systems characterization while assuming minimal background in math and physics.

- A broad audience including professionals across many fields, the general public and courses in colleges and even high schools.
- Wide STEM topics ranging from batteries to medicine with a writing style easy to follow.
- Short and inexpensive books of 120–150 pages main text that can be written and read in a reasonable amount of time.

Please contact the series editors, Bruce J. West (North Carolina State University, USA) and YangQuan Chen (University of California Merced), and Taylor & Francis Publisher Lian Sun (Lian.Sun@informa.com), if you have an idea for a book for the series.

Titles in the series currently include:

**Fractional Calculus for Skeptics I**
The Fractal Paradigm
*Bruce J. West and YangQuan Chen*

**On the Fractal Language of Medicine**
*Bruce J. West and W. Alan C. Mutch*

**Fractional Random Vibrations I**
Theories
*Ming Li*

**Fractional Random Vibrations II**
Applications
*Ming Li*

For more information about this book series, please visit https://www.routledge.com/Fractional-Order-Thinking-in-Exploring-the-Frontiers-of-STEM/book-series/FOT4STEM

# Fractional Random Vibrations II
## Applications

Ming Li

CRC Press
Taylor & Francis Group
Boca Raton London New York

CRC Press is an imprint of the
Taylor & Francis Group, an **informa** business

Designed cover image: © Ming Li

First edition published 2026
by CRC Press
2385 NW Executive Center Drive, Suite 320, Boca Raton FL 33431

and by CRC Press
4 Park Square, Milton Park, Abingdon, Oxon, OX14 4RN

*CRC Press is an imprint of Taylor & Francis Group, LLC*



ISBN: 978-1-041-11021-7 (hbk)
ISBN: 978-1-041-11022-4 (pbk)
ISBN: 978-1-003-65790-3 (ebk)

DOI: 10.1201/9781003657903

Typeset in Minion
by Apex CoVantage, LLC

*To my wife Yonglan Zhang, my daughter Joanna Jiayue Li, and my parents, Xidong Li and Fanggui Yin — for making it both possible and worthwhile*

# Contents

Preface to Volume II, xvi

Acknowledgements, xvii

CHAPTER 1 ▪ Responses of Fractional Vibrations Driven by Pierson-Moskowitz Spectrum 1

1.1 BACKGROUND 1

1.2 RESPONSES OF CLASS I FRACTIONAL VIBRATORS DRIVEN BY P-M SPECTRUM 2

1.2.1 Computations 2

1.2.2 Effect of $\alpha$ on Responses 5

1.3 RESPONSES OF CLASS II FRACTIONAL VIBRATORS DRIVEN BY P-M SPECTRUM 6

1.3.1 Computation Methods 6

1.3.2 Effect of $\beta$ on Responses 9

1.4 RESPONSES OF CLASS III FRACTIONAL VIBRATORS DRIVEN BY P-M SPECTRUM 11

1.4.1 Computations 11

1.4.2 Effect of $(\alpha, \beta)$ on Responses 13

1.5 RESPONSES OF CLASS IV FRACTIONAL VIBRATOR DRIVEN BY P-M SPECTRUM 14

1.5.1 Computations 14

1.5.2 Effect of $(\alpha, \lambda)$ on Responses 17

1.6 RESPONSES OF CLASS V FRACTIONAL VIBRATORS DRIVEN BY P-M SPECTRUM 18
1.6.1 Computations 18
1.6.2 Effect of $\lambda$ on Responses 21
1.7 RESPONSES OF CLASS VI FRACTIONAL VIBRATORS DRIVEN BY P-M SPECTRUM 21
1.7.1 Computations 21
1.7.2 Effect of $(\alpha, \beta, \lambda)$ on Responses 23
1.8 RESPONSES OF CLASS VII FRACTIONAL VIBRATORS DRIVEN BY P-M SPECTRUM 24
1.8.1 Computations 24
1.8.2 Effects of $\beta$, $\lambda$ on Responses 26
1.9 SUMMARY 27
1.10 EXERCISES 27
REFERENCES 28

CHAPTER 2 ▪ Responses of Fractional Vibrations Driven by Fractional Gaussian Noise 30

2.1 BACKGROUND 30
2.2 RESPONSES OF CLASS I FRACTIONAL VIBRATORS DRIVEN BY FGN 31
2.2.1 Computations 31
2.2.2 Effect of $\alpha$ on Responses 33
2.3 RESPONSES OF CLASS II FRACTIONAL VIBRATORS DRIVEN BY FGN 35
2.3.1 Computation Methods 35
2.3.2 Effect of $\beta$ on Responses 37
2.4 RESPONSES OF CLASS III FRACTIONAL VIBRATORS DRIVEN BY FGN 38
2.4.1 Computations 38
2.4.2 Effect of $(\alpha, \beta)$ on Responses 41
2.5 RESPONSES OF CLASS IV FRACTIONAL VIBRATORS DRIVEN BY FGN 42
2.5.1 Computations 42
2.5.2 Effect of $(\alpha, \lambda)$ on Responses 44

2.6 RESPONSES OF CLASS V FRACTIONAL VIBRATORS DRIVEN BY FGN 45
2.6.1 Computation Methods 45
2.6.2 Effect of $\lambda$ on Responses 47
2.7 RESPONSES OF CLASS VI FRACTIONAL VIBRATORS DRIVEN BY FGN 48
2.7.1 Computations 48
2.7.2 Effect of $(\alpha, \beta, \lambda)$ on Responses 50
2.8 RESPONSES OF CLASS VII FRACTIONAL VIBRATORS DRIVEN BY FGN 51
2.8.1 Computations 51
2.8.2 Effect of $(\beta, \lambda)$ on Responses 53
2.9 SUMMARY 54
2.10 EXERCISES 55
REFERENCES 56

CHAPTER 3 ▪ Responses of Fractional Vibrations Driven by Generalized Fractional Gaussian Noise 58

3.1 BACKGROUND 58
3.2 RESPONSES OF CLASS I FRACTIONAL VIBRATORS DRIVEN BY GFGN 59
3.2.1 Computations 59
3.2.2 Effect of $\alpha$ on Responses 61
3.3 RESPONSES OF CLASS II FRACTIONAL VIBRATORS DRIVEN BY GFGN 64
3.3.1 Computation Methods 64
3.3.2 Effect of $\beta$ on Responses 65
3.4 RESPONSES OF CLASS III FRACTIONAL VIBRATORS DRIVEN BY GFGN 67
3.4.1 Computations 67
3.4.2 Effect of $(\alpha, \beta)$ on Responses 69
3.5 RESPONSES OF CLASS IV FRACTIONAL VIBRATORS DRIVEN BY GFGN 69
3.5.1 Computations 69
3.5.2 Effect of $(\alpha, \lambda)$ on Responses 72

3.6 RESPONSES OF CLASS V FRACTIONAL VIBRATORS DRIVEN BY GFGN 73
3.6.1 Computation Methods 73
3.6.2 Effect of $\lambda$ on Responses 75
3.7 RESPONSES OF CLASS VI FRACTIONAL VIBRATORS DRIVEN BY GFGN 76
3.7.1 Computations 76
3.7.2 Effect of $(\alpha, \beta, \lambda)$ on Responses 78
3.8 RESPONSES OF CLASS VII FRACTIONAL VIBRATORS DRIVEN BY GFGN 78
3.8.1 Computations 78
3.8.2 Effect of $(\beta, \lambda)$ on Responses 81
3.9 SUMMARY 82
3.10 EXERCISES 82
REFERENCES 84

CHAPTER 4 ■ Responses of Fractional Vibrations Driven by Fractional Brownian Motion 85

4.1 BACKGROUND 85
4.2 RESPONSES OF CLASS I FRACTIONAL VIBRATORS DRIVEN BY FBM 86
4.2.1 Computations 86
4.2.2 Effect of $\alpha$ on Responses 89
4.3 RESPONSES OF CLASS II FRACTIONAL VIBRATORS DRIVEN BY FBM 90
4.3.1 Computation Methods 90
4.3.2 Effect of $\beta$ on Responses 94
4.4 RESPONSES OF CLASS III FRACTIONAL VIBRATORS DRIVEN BY FBM 97
4.4.1 Computations 97
4.4.2 Effect of $(\alpha, \beta)$ on Responses 99
4.5 RESPONSES OF CLASS IV FRACTIONAL VIBRATORS DRIVEN BY FBM 101
4.5.1 Computations 101
4.5.2 Effect of $(\alpha, \lambda)$ on Responses 103

4.6 RESPONSES OF CLASS V FRACTIONAL VIBRATORS DRIVEN BY FBM 103
4.6.1 Computation Methods 103
4.6.2 Effect of $\lambda$ on Responses 107
4.7 RESPONSES OF CLASS VI FRACTIONAL VIBRATORS DRIVEN BY FBM 107
4.7.1 Computations 107
4.7.2 Effect of ($\alpha$, $\beta$, $\lambda$) on Responses 110
4.8 RESPONSES OF CLASS VII FRACTIONAL VIBRATORS DRIVEN BY FBM 112
4.8.1 Computations 112
4.8.2 Effect of ($\beta$, $\lambda$) on Responses 114
4.9 SUMMARY 116
4.10 EXERCISES 116
REFERENCES 117

CHAPTER 5 ▪ Responses of Fractional Vibrations Driven by Fractional Ornstein-Uhlenbeck Processes 119

5.1 BACKGROUND 119
5.2 RESPONSES OF CLASS I FRACTIONAL VIBRATORS DRIVEN BY FRACTIONAL OU PROCESSES 120
5.2.1 Computations 120
5.2.2 Effect of $\alpha$ on Responses 122
5.3 RESPONSES OF CLASS II FRACTIONAL VIBRATION SYSTEMS DRIVEN BY FRACTIONAL OU PROCESSES 124
5.3.1 Computation Methods 124
5.3.2 Effect of $\beta$ on Responses 127
5.4 RESPONSES OF CLASS III FRACTIONAL VIBRATORS DRIVEN BY FRACTIONAL OU PROCESSES 129
5.4.1 Computations 129
5.4.2 Effect of ($\alpha$, $\beta$) on Responses 131
5.5 RESPONSES OF CLASS IV FRACTIONAL VIBRATION SYSTEMS DRIVEN BY FRACTIONAL OU PROCESSES 132
5.5.1 Computations 132
5.5.2 Effect of ($\alpha$, $\lambda$) on Responses 134

5.6 RESPONSES OF CLASS V FRACTIONAL VIBRATORS DRIVEN BY FRACTIONAL OU PROCESSES 135
5.6.1 Computation Methods 135
5.6.2 Effect of $\lambda$ on Responses 137
5.7 RESPONSES OF CLASS VI FRACTIONAL VIBRATORS DRIVEN BY FRACTIONAL OU PROCESSES 138
5.7.1 Computations 138
5.7.2 Effect of ($\alpha$, $\beta$, $\lambda$) on Responses 140
5.8 RESPONSES OF CLASS VII FRACTIONAL VIBRATORS DRIVEN BY FRACTIONAL OU PROCESSES 141
5.8.1 Computations 141
5.8.2 Effect of ($\beta$, $\lambda$) on Responses 143
5.9 SUMMARY 144
5.10 EXERCISES 145
REFERENCES 145

CHAPTER 6 ■ Responses of Fractional Vibrations Driven by von Kármán Spectrum 147

6.1 BACKGROUND 147
6.2 RESPONSES OF CLASS I FRACTIONAL VIBRATORS DRIVEN BY VON KÁRMÁN SPECTRUM 148
6.2.1 Computations 148
6.2.2 Effect of $\alpha$ on Responses 150
6.3 RESPONSES OF CLASS II FRACTIONAL VIBRATION SYSTEMS DRIVEN BY VON KÁRMÁN SPECTRUM 152
6.3.1 Computation Methods 152
6.3.2 Effect of $\beta$ on Responses 155
6.4 RESPONSES OF CLASS III FRACTIONAL VIBRATORS DRIVEN BY VON KÁRMÁN SPECTRUM 156
6.4.1 Computations 156
6.4.2 Effect of ($\alpha$, $\beta$) on Responses 158
6.5 RESPONSES OF CLASS IV FRACTIONAL VIBRATION SYSTEMS DRIVEN BY VON KÁRMÁN SPECTRUM 158
6.5.1 Computations 158
6.5.2 Effect of ($\alpha$, $\lambda$) on Responses 162

6.6 RESPONSES OF CLASS V FRACTIONAL VIBRATORS DRIVEN BY VON KÁRMÁN SPECTRUM 163
6.6.1 Computation Methods 163
6.6.2 Effect of $\lambda$ on Responses 165
6.7 RESPONSES OF CLASS VI FRACTIONAL VIBRATORS DRIVEN BY VON KÁRMÁN SPECTRUM 165
6.7.1 Computations 165
6.7.2 Effect of ($\alpha$, $\beta$, $\lambda$) on Responses 168
6.8 RESPONSES OF CLASS VII FRACTIONAL VIBRATORS DRIVEN BY VON KÁRMÁN SPECTRUM 169
6.8.1 Computations 169
6.8.2 Effect of ($\beta$, $\lambda$) on Responses 171
6.9 SUMMARY 171
6.10 EXERCISES 172
REFERENCES 173

CHAPTER 7 ▪ Postscript to Volume II 174

INDEX, 176

# Preface to Volume II

VOLUME II CONSISTS OF SEVEN chapters with respect to the applications of the theories of seven classes of fractional vibration systems addressed in Volume I. Chapter 1 presents the closed-form expressions of power spectrum density (PSD) response and cross-PSD response of seven classes of fractional vibrators driven by fully developed ocean surface waves with the Pierson-Moskowitz spectrum, Chapter 2 proposes the closed-form expressions of PSD response and cross-PSD response of seven classes of fractional vibrators driven by fractional Gaussian noise, Chapter 3 gives the closed-form expressions of PSD response and cross-PSD response of seven classes of fractional vibrators under the excitation of generalized fractional Gaussian noise, Chapter 4 brings forward the closed-form expressions of PSD response and cross-PSD response of seven classes of fractional vibrators driven by fractional Brownian motion, Chapter 5 puts forward the closed form expressions of PSD response and cross-PSD response of seven classes of fractional vibrators driven by fractional Ornstein-Uhlenbeck process, Chapter 6 establishes the closed-form expressions of PSD response and cross-PSD one of seven classes of fractional vibrators excited by wind fluctuation speed with the von Kármán spectrum, and finally Chapter 7 is the postscript to Volume II. Excitations of the Pierson-Moskowitz spectrum and the von Kármán one are widely used in marine engineering and wind engineering, while other fractional processes may be quite academic at present. For all chapters, one thing in common is that there are considerable effects of fractional order(s) of seven classes of fractional vibration systems on responses.

**Ming Li, Ph.D., Professor, Grand Secretary Li XVI (Rizhao)**
*Ocean College, Zhejiang University, PR. China*

# Acknowledgements

Thanks go to Prof. Jianxing Leng and Ocean College, Zhejiang University, China, for providing me with the professor position to lecture on the course of ship hull vibrations for undergraduates and postgraduates. Without my teaching in Zhejiang University, it would have been impossible for me to arrange my time to write this monograph. Prof. Xuekang Gu, the vice technical director of China Ship Scientific Research Center (CSSRC); Prof. Yousheng Wu (Ex-Director of the CSSRC, the academician of the Chinese Academy of Engineering); and Prof. Jingjian Chen (CSSRC) are acknowledged.

I would like to take the opportunity to acknowledge a group of professors in Tsinghua University (Beijing) for their education in hard times. They are Qiji Yang, Shiliang Xu, Desheng Wang, Deyun Lin, Dingyue Kou, Baoqin Liu, Jingzhao She, Daimao Lin, Jingxian Zou, Minsheng Hua, Xiaoqing Ding, Xiguang Ma, Xueli Qiao, Xuexia Zhang, Shichang Hou, Zhenming Feng, Jiaguang Fang, Jiaqing Li, Bo Lv, Siming Luo, Guoxiang Zhao, Mengtao Wang, and Xiyuan Yan.

Prof. YangQuan Chen (University of California, Merced) and Prof. Bruce J. West (US Army Research Office, University of Rochester) for their instructions, comments, and encouragement on this monograph are particularly appreciated. Prof. Swee Cheng Lim (Multimedia University) is appreciated for his instructions and discussions in fractional processes. I am grateful to Mr. Yong Chen for improving the resolution of some drawings. Thanks forever go to my uncle (Kerui Li) and aunt (Zhijian Fang) for their love and encouragement.

The views and conclusions contained in this book are those of the author and should not be interpreted as representing the official policies, either expressed or implied, of the Chinese government.

CHAPTER 1

# Responses of Fractional Vibrations Driven by Pierson-Moskowitz Spectrum

THE CONTRIBUTIONS GIVEN IN this chapter are twofold. One is to propose the analytic expressions of the power spectrum density (PSD) responses and cross-PSD responses to fractional vibrators from class I to class VII under the excitation of the Pierson and Moskowitz (P-M) spectrum using elementary functions. The other is to show that the fractional orders, namely, $\alpha$, $\beta$, and $\lambda$, have considerable effects on responses when ocean surface waves pass through those fractional vibration systems.

## 1.1 BACKGROUND

Recently, the analytic theory of seven classes of fractional vibration systems using elementary functions was established by Li ([1, 2]), also see Chapter 7 in Volume I. In the field of ship science and technology, an interesting topic is about responses of structural vibrations driven by fully developed ocean surface waves (Jensen [3]). However, reports regarding the responses of seven classes of fractional vibration systems addressed in Li [1, 2] and Chapter 7 in Volume I, which are driven by the Pierson and

DOI: 10.1201/9781003657903-1

Moskowitz (P-M) spectrum, are rarely seen. The aim of this chapter is to present the analytic theory of the power spectrum density (PSD) responses and cross-PSD ones of seven classes of fractional vibration systems driven by the P-M spectrum.

There are several spectral models of fully developed ocean surface waves, see, for example, Okumoto et al. [4], Pierson and Moskowitz [5], Hasselmann et al. [6], Ochi and Hubble [7], Scott [8], and Phillips [9]. Among them, the spectrum reported by Pierson and Moskowitz [5] is widely used as a reference standard regarding spectral models of fully developed ocean surface waves (Jensen [3], Massel [10], Chairabarti [11], Li [12], and ITTC [13]). Thus, we take the P-M spectrum as a representative of fully developed ocean surface waves in this chapter.

The rest of the chapter is organized as follows. The PSD responses and cross-PSD responses to seven classes of fractional vibration systems under the excitation of the P-M spectrum are put forward in Sections 1.2–1.8, respectively. The summary is given in Section 1.9.

## 1.2 RESPONSES OF CLASS I FRACTIONAL VIBRATORS DRIVEN BY P-M SPECTRUM

### 1.2.1 Computations

Consider the motion equation of a class I fractional vibrator in the form

$$m\frac{d^{\alpha}x_1(t)}{dt^{\alpha}}+k\frac{dx_1(t)}{dt}=f(t). \tag{1.1}$$

In (1.1), $1<\alpha<3$, $x_1(t)$ is the response, $f(t)$ is driven force, $m$ and $k$ are the primary mass and stiffness, respectively.

Let $f(t)$ be the driven force with the P-M spectrum. Let $S_{ff}(\omega)$ be the PSD of $f(t)$. Then, $S_{ff}(\omega)$ is given by

$$S_{ff}(\omega)=ag^2\omega^{-5}e^{-b\left(\frac{g}{V}\right)^4\frac{1}{\omega^4}}. \tag{1.2}$$

In (1.2), $a=8.1\times10^{-3}$, $b=0.74$, $g$ is the acceleration of gravity (m/s$^2$), and $V$ wind speed (m/s) at an elevation of 19.5 m above the sea surface. The unit of $S$ is m$^2\cdot$s. Figure 1.1 indicates a plot of $S_{ff}(\omega)$.

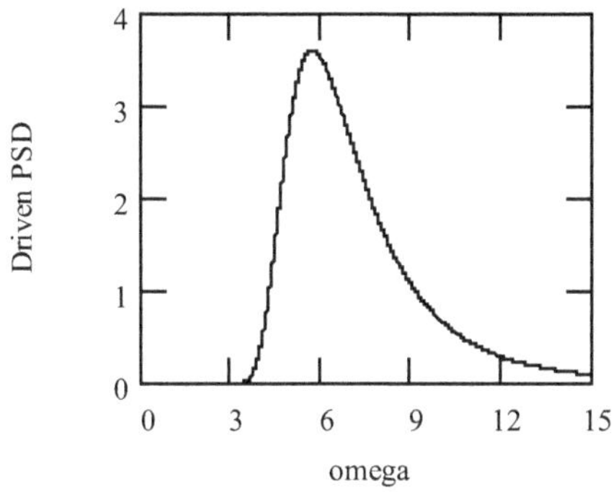


FIGURE 1.1 Plot of P-M spectrum with $V = 15$.

**Theorem 1.1 (PSD response I)**

Denote by $S_{xx1}(\omega)$ the PSD of $x_1(t)$. Then,

$$S_{xx1}(\omega) = \frac{ag^2\omega^{-5}e^{-b\left(\frac{g}{V}\right)^4\frac{1}{\omega^4}}}{k^2\left[\left(1-\frac{\omega^\alpha}{\omega_n^2}\left|\cos\frac{\alpha\pi}{2}\right|\right)^2+\left(\frac{\omega^\alpha}{\omega_n^2}\sin\frac{\alpha\pi}{2}\right)^2\right]}. \tag{1.3}$$

In (1.3), $\omega_n^2 = \frac{k}{m}$.

*Proof.* According to Chapter 4 in Volume I, $S_{xx1}(\omega) = S_{ff}(\omega)|H_1(\omega)|^2$. Following Chapter 7 in Volume I or Li [1] or Li [2], we have

$$|H_1(\omega)| = \frac{1/k}{\sqrt{\left(1-\frac{\omega^\alpha}{\omega_n^2}\left|\cos\frac{\alpha\pi}{2}\right|\right)^2+\left(\frac{\omega^\alpha}{\omega_n^2}\sin\frac{\alpha\pi}{2}\right)^2}}.$$

Thus, (1.13) holds. The proof is finished.

**Theorem 1.2 (cross-PSD response I)**

Let $S_{fx1}(\omega)$ be the cross PSD between $f(t)$ and $x_1(t)$. Then,

$$S_{fx1}(\omega)=\frac{ag^2\omega^{-5}e^{-b\left(\frac{g}{V}\right)^4\frac{1}{\omega^4}}}{k\left(1-\frac{\omega^\alpha}{\omega_n^2}\left|\cos\frac{\alpha\pi}{2}\right|+i\frac{\omega^\alpha}{\omega_n^2}\sin\frac{\alpha\pi}{2}\right)}. \tag{1.4}$$

*Proof.* According to Chapter 7 in Volume I or Li [1] or [2], $H_1(\omega)$ is expressed by

$$H_1(\omega)=\frac{1}{k\left(1-\frac{\omega^\alpha}{\omega_n^2}\left|\cos\frac{\alpha\pi}{2}\right|+i\frac{\omega^\alpha}{\omega_n^2}\sin\frac{\alpha\pi}{2}\right)}. \tag{1.5}$$

According to Chapter 4 in Volume I, $S_{fx1}(\omega) = S_{ff}(\omega)H_1(\omega)$. From (1.2) and (1.5), we see that (1.4) is sound. This finishes the proof.

Let $r_{ff}(\tau)$ be the autocorrelation function (ACF) of $f(t)$. Let $h_1(t)$ be the impulse response of a class I fractional vibrator. Denote by $r_{xx1}(\tau)$ the ACF response, that is, the ACF of $x_1(t)$. In the time domain, therefore, we have

$$r_{xx1}(\tau) = r_{ff}(\tau)*h_1(\tau)*h_1(-\tau). \tag{1.6}$$

In (1.6), $*$ stands for the convolution operation. Denote by $r_{fx1}(\tau)$ the cross-correlation response between $f(t)$ and $x_1(t)$. Then,

$$r_{fx1}(\tau) = r_{ff}(\tau)*h_1(\tau). \tag{1.7}$$

According to Chapter 7 in Volume I or Li [1] or [2],

$$h_1(t)=\frac{e^{-\frac{\omega\sin\frac{\alpha\pi}{2}}{2\left|\cos\frac{\alpha\pi}{2}\right|}t}\sin\left(\frac{\omega_n}{\sqrt{\omega^{\alpha-2}\left|\cos\frac{\alpha\pi}{2}\right|}}\sqrt{1-\frac{\omega^{2\alpha}\sin^2\frac{\alpha\pi}{2}}{4\omega_n^2\left|\cos\frac{\alpha\pi}{2}\right|}}\,t\right)}{m\omega_n\sqrt{\omega^{\alpha-2}\left|\cos\frac{\alpha\pi}{2}\right|}\sqrt{1-\frac{\omega^{2\alpha}\sin^2\frac{\alpha\pi}{2}}{4\omega_n^2\left|\cos\frac{\alpha\pi}{2}\right|}}}u(t). \tag{1.8}$$

In (1.8), $u(t)$ is the unit step function.

### 1.2.2 Effect of $\alpha$ on Responses

Figure 1.2 indicates some plots of $|H_1(\omega)|^2$. Some plots of $S_{xx1}(\omega)$ are shown in Figure 1.3.

When $\alpha = 2$, a class I fractional vibrator reduces to be a conventional damping-free vibrator. For $\alpha = 2$, $m = 1$, $c = 0.1$, $k = 36$, and $V = 15$, a resonance occurs as indicated in Figure 1.4. In Figure 1.5, we illustrate some plots of the cross-PSD response $S_{fx1}(\omega)$.

According to the random data generation used in Massel [10], and Chairabarti [11], Li [14–17], we illustrate some plots of driven signal $f(t)$ and the response $x_1(t)$ in Figure 1.6. From Figures 1.3–1.6, we see that there is noticeable effect of the fractional order $\alpha$ on the responses of class I fractional vibration systems driven by the P-M spectrum.

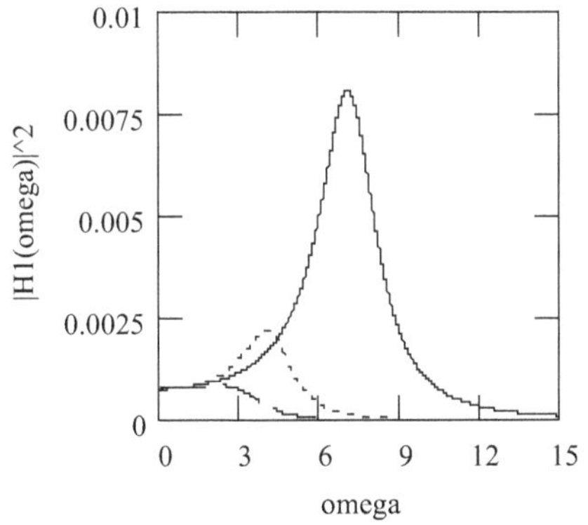


FIGURE 1.2 Plots of $|H_1(\omega)|^2$ for $\alpha = 1.8$ (solid), 2.4 (dot), 2.8 (dash) when $m = 1$ and $k = 36$.

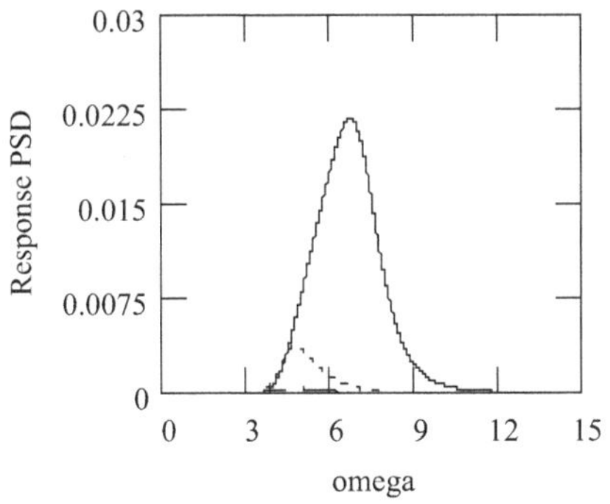


FIGURE 1.3 Plots of $S_{xx1}(\omega)$ for $\alpha = 1.8$ (solid), 2.4 (dot), 2.8 (dash) when $m = 1$, $k = 36$, and $V = 15$.

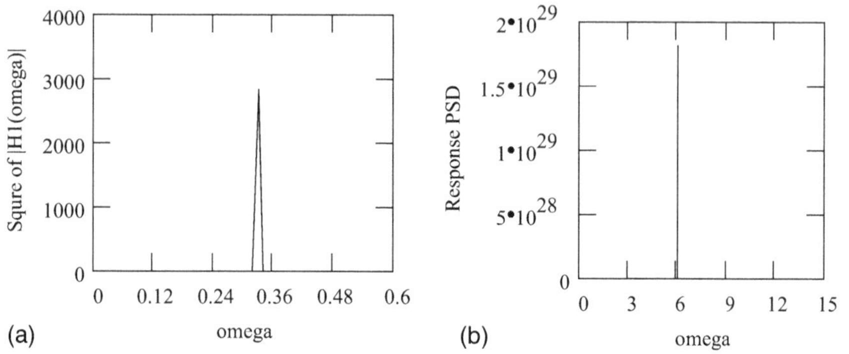


FIGURE 1.4 Resonance when $\alpha = 2$, $m = 1$, $k = 36$, and $V = 15$. (a). Plot of $|H_1(\omega)|^2$. (b). Plot of response PSD $S_{xx1}(\omega)$.

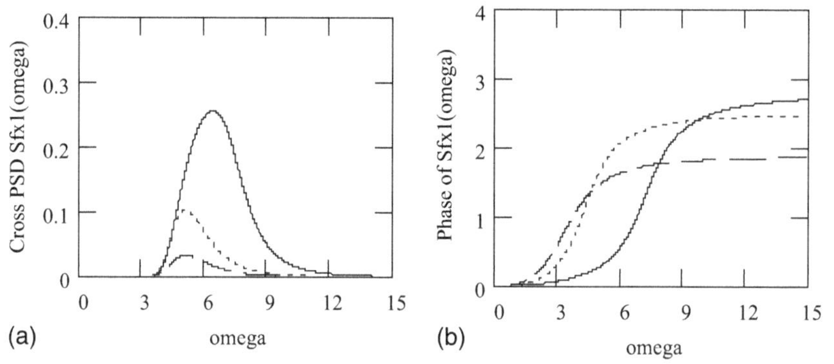


FIGURE 1.5 Illustrations of $S_{fx1}(\omega)$ for $\alpha = 1.8$ (solid), 2.4 (dot), 2.8 (dash) when $m = 1$, $k = 36$, and $V = 15$. (a). $|S_{fx1}(\omega)|$. (b). Phase of $S_{fx1}(\omega)$.

## 1.3 RESPONSES OF CLASS II FRACTIONAL VIBRATORS DRIVEN BY P-M SPECTRUM

### 1.3.1 Computation Methods

For a class II fractional vibrator, its motion equation is given by

$$m\frac{d^2x_2(t)}{dt^2} + c\frac{d^\beta x_2(t)}{dt^\beta} + k\frac{dx_2(t)}{dt} = f(t). \tag{1.9}$$

In (1.9), $0 < \beta < 2$, $x_2(t)$ is the response and $c$ is the primary damping.

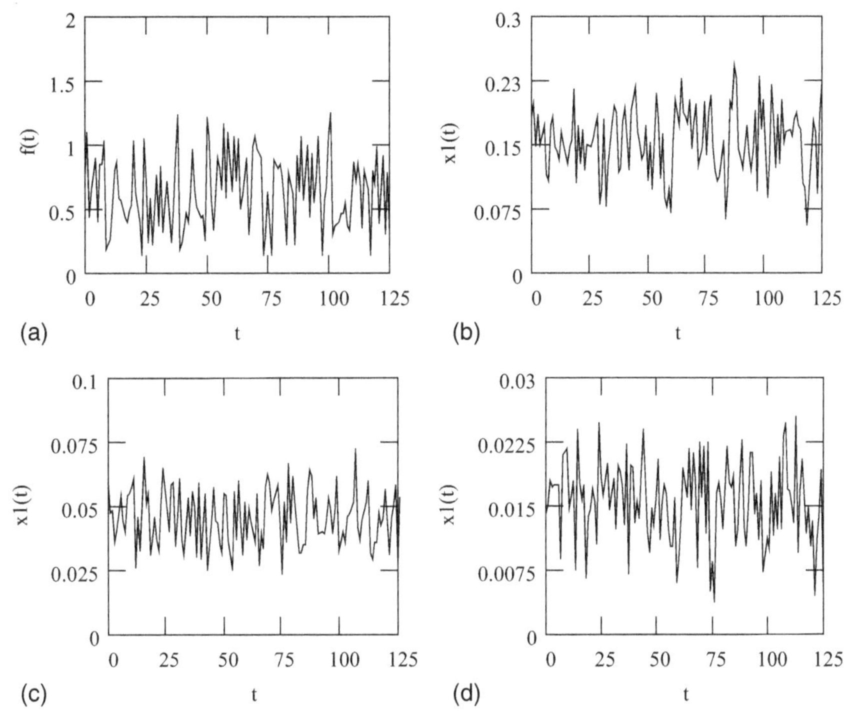


FIGURE 1.6 Plots of driven signal and response when $m = 1$, $k = 36$, and $V = 15$. (a). Driven signal $f(t)$. (b). Response $x_1(t)$ for $\alpha = 1.8$. (c). Response $x_1(t)$ for $\alpha = 2.4$. (d). Response $x_1(t)$ for $\alpha = 2.8$.

**Theorem 1.3 (PSD response II)**

Let $S_{xx2}(\omega)$ be the PSD of $x_2(t)$. Then,

$$S_{xx2}(\omega) = \frac{a g^2 \omega^{-5} e^{-b\left(\frac{g}{V}\right)^4 \frac{1}{\omega^4}}}{k^2 \left\{ \left[ 1 - \gamma^2 \left( 1 - \frac{c}{m} \omega^{\beta-2} \cos\frac{\beta\pi}{2} \right) \right]^2 + \left( \frac{2\varsigma\omega^{\beta}}{\omega_n} \sin\frac{\beta\pi}{2} \right)^2 \right\}}. \tag{1.10}$$

In (1.10), $\gamma = \dfrac{\omega}{\omega_n}$.

*Proof.* Note that $S_{xx2}(\omega) = S_{ff}(\omega)|H_2(\omega)|^2$ (Li [17, 18]), where $|H_2(\omega)|$ is given by

$$|H_2(\omega)| = \frac{1/k}{\sqrt{\left[1-\gamma^2\left(1-\frac{c}{m}\omega^{\beta-2}\cos\frac{\beta\pi}{2}\right)\right]^2+\left(\frac{2\varsigma\omega^{\beta}}{\omega_n}\sin\frac{\beta\pi}{2}\right)^2}}. \tag{1.11}$$

Thus, Theorem 1.3 is valid. This finishes the proof.

**Theorem 1.4 (cross-PSD response II)**

Denote by $S_{fx2}(\omega)$ the cross PSD between $f(t)$ and $x_2(t)$. Then,

$$S_{fx2}(\omega) = \frac{ag^2\omega^{-5}e^{-b\left(\frac{g}{V}\right)^4\frac{1}{\omega^4}}}{k\left[1-\gamma^2\left(1-2\varsigma\omega_n\omega^{\beta-2}\cos\frac{\beta\pi}{2}\right)+i\frac{2\varsigma\omega^{\beta}}{\omega_n}\sin\frac{\beta\pi}{2}\right]}. \tag{1.12}$$

*Proof.* Due to $S_{fx2}(\omega) = S_{ff}(\omega)H_2(\omega)$, where

$$H_2(\omega) = \frac{1/k}{1-\gamma^2\left(1-2\varsigma\omega_n\omega^{\beta-2}\cos\frac{\beta\pi}{2}\right)+i\frac{2\varsigma\omega^{\beta}\sin\frac{\beta\pi}{2}}{\omega_n}},$$

(1.12) holds. The proof completes.

Let $r_{xx2}(\tau)$ be the ACF of $x_2(t)$. Denote by $h_2(\tau)$ the impulse response of a class II fractional vibrator. Then,

$$r_{xx2}(\tau) = r_{ff}(\tau) * h_2(\tau) * h_2(-\tau). \tag{1.13}$$

Let $r_{fx2}(\tau)$ be the cross-correlation between $f(t)$ and $x_2(t)$. Then,

$$r_{fx2}(\tau) = r_{ff}(\tau) * h_2(\tau). \tag{1.14}$$

In (1.13) and (1.14) (Chapter 7 in Volume I or Li [1] or [2]), $h_2(t)$ is given by

$$h_2(t) = \frac{e^{-\frac{\varsigma\omega_n\omega^{\beta-1}\sin\frac{\beta\pi}{2}}{1-\frac{c}{m}\omega^{\beta-2}\cos\frac{\beta\pi}{2}}t}\sin\frac{\omega_n\sqrt{1-\frac{\varsigma^2\omega^{2(\beta-1)}\sin^2\frac{\beta\pi}{2}}{1-\frac{c}{m}\omega^{\beta-2}\cos\frac{\beta\pi}{2}}}}{\sqrt{1-\frac{c}{m}\omega^{\beta-2}\cos\frac{\beta\pi}{2}}}t}{\omega_n m\sqrt{1-\frac{c}{m}\omega^{\beta-2}\cos\frac{\beta\pi}{2}}\sqrt{1-\frac{\varsigma^2\omega^{2(\beta-1)}\sin^2\frac{\beta\pi}{2}}{1-\frac{c}{m}\omega^{\beta-2}\cos\frac{\beta\pi}{2}}}}u(t). \tag{1.15}$$

### 1.3.2 Effect of $\beta$ on Responses

Figure 1.7 shows some plots of $|H_2(\omega)|^2$. Figure 1.8 indicates some plots of $S_{xx2}(\omega)$. Figure 1.9 shows resonance curves when $\beta = 1$. Figure 1.10 illustrates some plots of $S_{fx2}(\omega)$. Figure 1.11 exhibits the effects of $\beta$ on the fluctuation range of the response $x_2(t)$. As can be seen from Figures 1.8–1.11, there is effect of $\beta$ on the responses to class II fractional vibration systems.

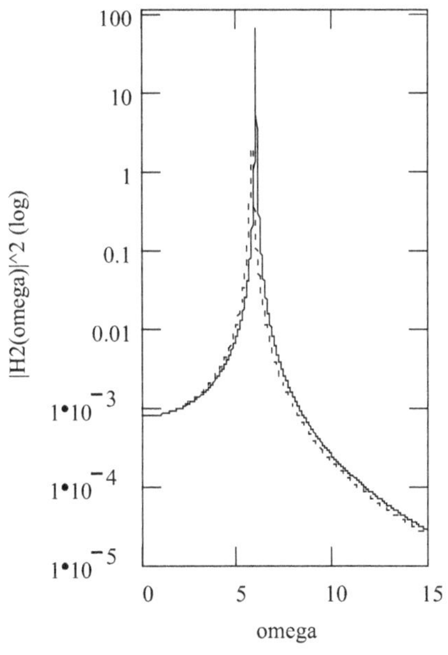


FIGURE 1.7 Plots of $|H_2(\omega)|^2$ (log) with $\beta = 0.4$ (solid), 1.8 (dot), when $m = 1$, $c = 0.1$, and $k = 36$.

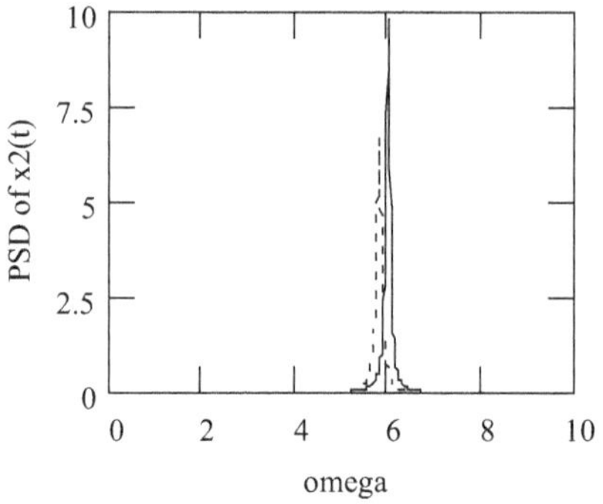


FIGURE 1.8 Plots of $S_{xx2}(\omega)$ with $\beta = 0.4$ (solid), 1.8 (dot), when $m = 1$, $c = 0.1$, $k = 36$, and $V = 15$.

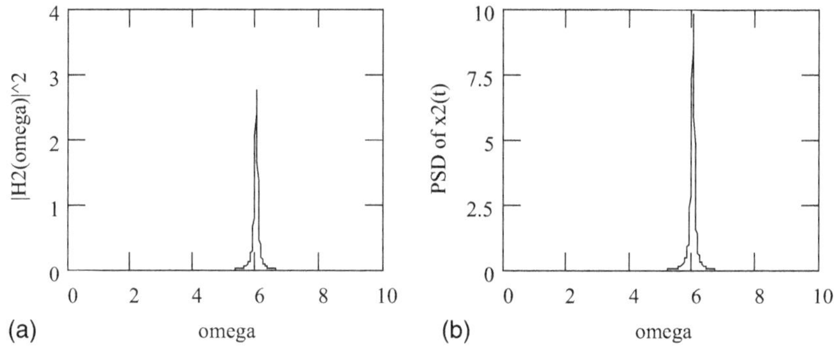


FIGURE 1.9 Resonance when $\beta = 1$, $m = 1$, $c = 0.1$, $k = 36$, and $V = 15$. (a). $|H_2(\omega)|^2$. (b). PSD $S_{xx2}(\omega)$.

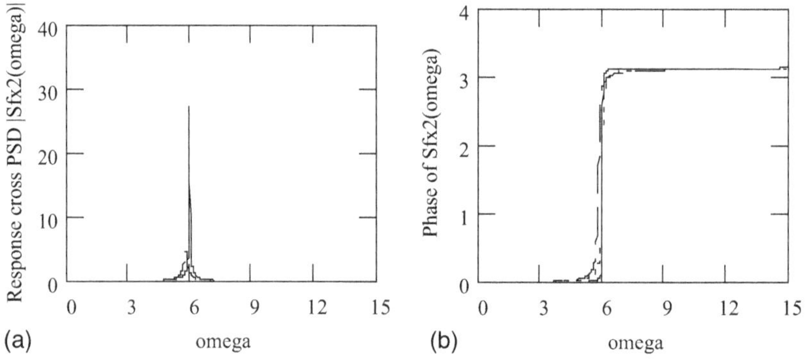


FIGURE 1.10 Plots of $S_{fx2}(\omega)$ with $\beta = 0.4$ (solid), 1 (dot), 1.8 (dash), when $m = 1$, $c = 0.1$, $k = 36$, and $V = 15$. (a). $|S_{fx2}(\omega)|$. (b). Phase of $S_{fx2}(\omega)$.

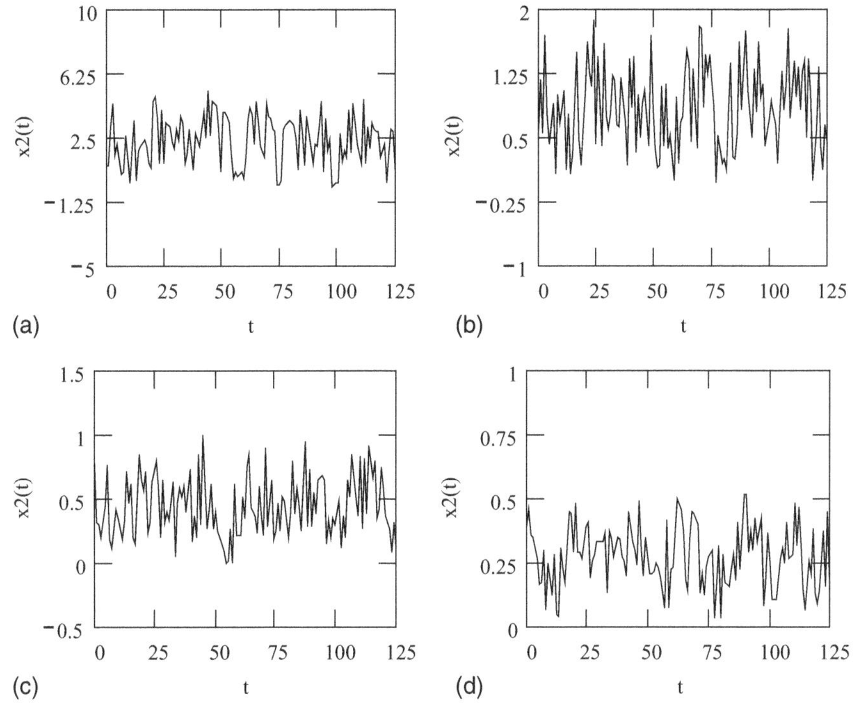


FIGURE 1.11 Simulated response series $x_2(t)$ when $m = 1$, $c = 0.1$, $k = 36$, and $V = 15$. (a). $x_2(t)$ for $\beta = 0.4$. (b). $x_2(t)$ for $\beta = 1$. (c). $x_2(t)$ for $\beta = 1.4$. (d). $x_2(t)$ for $\beta = 1.8$.

## 1.4 RESPONSES OF CLASS III FRACTIONAL VIBRATORS DRIVEN BY P-M SPECTRUM

### 1.4.1 Computations

The motion equation of a class III fractional vibrator is given by

$$m\frac{d^{\alpha}x_3(t)}{dt^{\alpha}}+c\frac{d^{\beta}x_3(t)}{dt^{\beta}}+kx_3(t)=f(t). \tag{1.16}$$

In (1.16), $x_3(t)$ is the response of a class III fractional vibrator.

**Theorem 1.5 (PSD response III)**

Let $S_{xx3}(\omega)$ be the PSD of $x_3(t)$. Then,

$$S_{xx3}(\omega)=\frac{ag^2\omega^{-5}e^{-b\left(\frac{g}{V}\right)^4\frac{1}{\omega^4}}}{k^2\left\{\left[1-\gamma^2\left(\omega^{\alpha-2}\left|\cos\frac{\alpha\pi}{2}\right|-2\varsigma\omega_n\omega^{\beta-2}\cos\frac{\beta\pi}{2}\right)\right]^2+\left[\frac{\gamma\left(\omega^{\alpha-1}\sin\frac{\alpha\pi}{2}+2\varsigma\omega_n\omega^{\beta-1}\sin\frac{\beta\pi}{2}\right)}{\omega_n\left(\omega^{\alpha-2}\left|\cos\frac{\alpha\pi}{2}\right|-2\varsigma\omega_n\omega^{\beta-2}\cos\frac{\beta\pi}{2}\right)}\right]^2\right\}}. \quad (1.17)$$

*Proof.* Note that $S_{xx3}(\omega) = S_{ff}(\omega)|H_3(\omega)|^2$. According to Chapter 7 in Volume I or Li [1] or [2], $H_3(\omega)$ is expressed by

$$H_3(\omega)=\frac{1/k}{1-\gamma^2\left(\omega^{\alpha-2}\left|\cos\frac{\alpha\pi}{2}\right|-2\varsigma\omega_n\omega^{\beta-2}\cos\frac{\beta\pi}{2}\right)+i\frac{\gamma\left(\omega^{\alpha-1}\sin\frac{\alpha\pi}{2}+2\varsigma\omega_n\omega^{\beta-1}\sin\frac{\beta\pi}{2}\right)}{\omega_n\left(\omega^{\alpha-2}\left|\cos\frac{\alpha\pi}{2}\right|-2\varsigma\omega_n\omega^{\beta-2}\cos\frac{\beta\pi}{2}\right)}}. \quad (1.18)$$

Thus, (1.17) holds. The proof is finished.

**Theorem 1.6 (cross-PSD response III)**

Let $S_{fx3}(\omega)$ be the cross PSD between $f(t)$ and $x_3(t)$. Then,

$$S_{fx3}(\omega)=\frac{k^{-1}ag^2\omega^{-5}e^{-b\left(\frac{g}{V}\right)^4\frac{1}{\omega^4}}}{1-\gamma^2\left(\omega^{\alpha-2}\left|\cos\frac{\alpha\pi}{2}\right|-2\varsigma\omega_n\omega^{\beta-2}\cos\frac{\beta\pi}{2}\right)+i\frac{\gamma\left(\omega^{\alpha-1}\sin\frac{\alpha\pi}{2}+2\varsigma\omega_n\omega^{\beta-1}\sin\frac{\beta\pi}{2}\right)}{\omega_n\left(\omega^{\alpha-2}\left|\cos\frac{\alpha\pi}{2}\right|-2\varsigma\omega_n\omega^{\beta-2}\cos\frac{\beta\pi}{2}\right)}}. \quad (1.19)$$

*Proof.* As $S_{fx3}(\omega) = S_{ff}(\omega)H_3(\omega)$ and considering (1.18), we have (1.19). The proof completes.

Denote by $r_{xx3}(\tau)$ the ACF of response $x_3(t)$. Let $h_3(\tau)$ be the impulse response of a class III fractional vibrator. Then,

$$r_{xx3}(\tau) = r_{ff}(\tau) * h_3(\tau) * h_3(-\tau). \tag{1.20}$$

Let $r_{fx3}(\tau)$ be the cross-correlation response between $f(t)$ and $x_3(t)$. Then,

$$r_{fx3}(\tau) = r_{ff}(\tau) * h_3(\tau). \tag{1.21}$$

In (1.20) and (1.21), $h_3(\tau)$ is given by

$$h_3(t) = \frac{e^{-\frac{m\omega^{\alpha-1}\sin\frac{\alpha\pi}{2}+c\omega^{\beta-1}\sin\frac{\beta\pi}{2}}{2\sqrt{-\left(m\omega^{\alpha-2}\cos\frac{\alpha\pi}{2}+c\omega^{\beta-2}\cos\frac{\beta\pi}{2}\right)k}}\omega_{\text{eqn3}}t}\sin\omega_{\text{eqd3}}t}{-\left(m\omega^{\alpha-2}\cos\frac{\alpha\pi}{2}+c\omega^{\beta-2}\cos\frac{\beta\pi}{2}\right)\omega_{\text{eqd3}}}u(t). \tag{1.22}$$

Refer to Chapter 7 in Volume I or Li [1] or [2] for the expressions of $\omega_{\text{eqn3}}$ and $\omega_{\text{eqd3}}$ in (1.22).

### 1.4.2 Effect of ($\alpha$, $\beta$) on Responses

We illustrate some plots of $|H_3(\omega)|^2$ in Figure 1.12. Figure 1.13 indicates some plots of $S_{xx3}(\omega)$. Some plots of $S_{fx3}(\omega)$ are indicated in Figure 1.14. Figures 1.13 and 1.14 show that the effect of the fractional orders ($\alpha$, $\beta$) on the responses of class III fractional vibrators is significant.

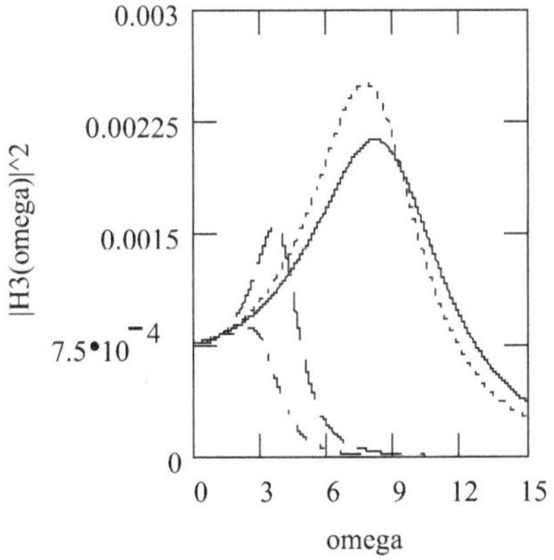


FIGURE 1.12 Plots of $|H_3(\omega)|^2$ when $m = 1$, $c = 0.1$, and $k = 36$, for $(\alpha, \beta) = (1.6, 0.8)$ (solid), (1.6, 1.8) (dot), (2.5, 0.8) (dash), (2.8, 1.8) (dash dot).

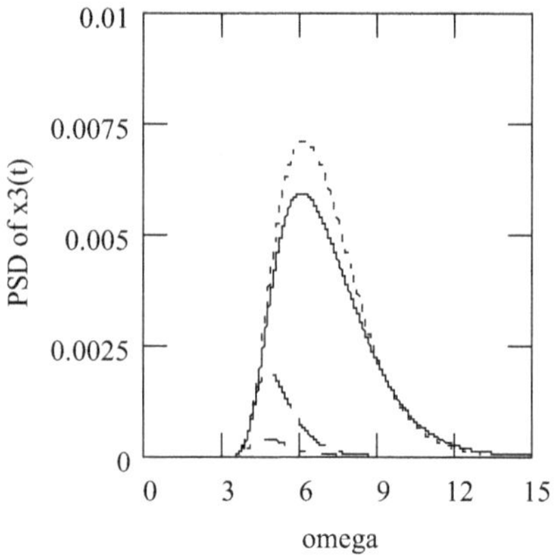


FIGURE 1.13 Plots of $S_{xx3}(\omega)$ when $m = 1$, $c = 0.1$, $k = 36$, and $V = 15$ for $(\alpha, \beta) =$ (1.6, 0.8) (solid), (1.6, 1.8) (dot), (2.5, 0.8) (dash), (2.8, 1.8) (dash dot).

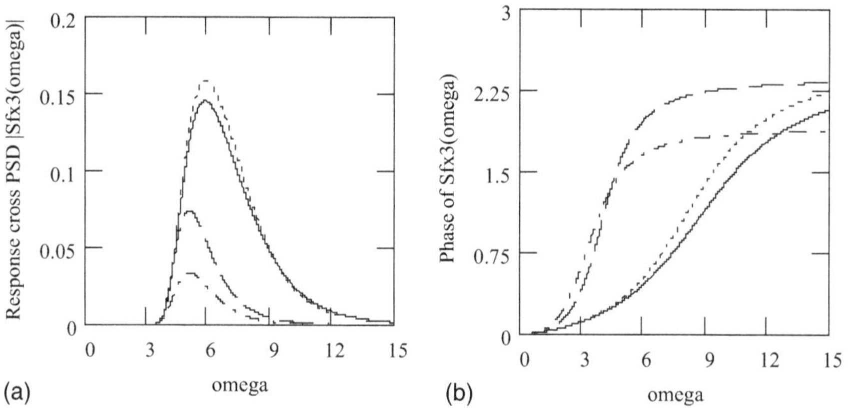


FIGURE 1.14 Plots of $S_{fx3}(\omega)$ when $m = 1$, $c = 0.1$, $k = 36$, and $V = 15$ for $(\alpha, \beta) =$ (1.6, 0.8) (solid), (1.6, 1.8) (dot), (2.5, 0.8) (dash), (2.8, 1.8) (dash dot). (a). $|S_{fx3}(\omega)|$. (b). Phase of $S_{fx3}(\omega)$.

## 1.5 RESPONSES OF CLASS IV FRACTIONAL VIBRATOR DRIVEN BY P-M SPECTRUM

### 1.5.1 Computations

The motion equation of a class IV fractional vibrator is given by

$$m\frac{d^{\alpha}x_4(t)}{dt^{\alpha}} + k\frac{d^{\lambda}x_4(t)}{dt^{\lambda}} = f(t). \tag{1.23}$$

In (1.23), $0 \le \lambda < 1$, $x_4(t)$ is the response of a class IV fractional vibrator.

**Theorem 1.7 (PSD response IV)**

Let $S_{xx4}(\omega)$ be the PSD of $x_4(t)$. Then,

$$S_{xx4}(\omega)=\frac{ag^2\omega^{-5}e^{-b\left(\frac{g}{V}\right)^4\frac{1}{\omega^4}}}{k^2\omega^{2\lambda}\cos^2\frac{\lambda\pi}{2}\left[\left(1-\gamma^2\frac{-\omega^{\alpha-2}\cos\frac{\alpha\pi}{2}}{\omega^{\lambda}\cos\frac{\lambda\pi}{2}}\right)^2+4\left(\gamma\frac{m\omega^{\alpha-1}\sin\frac{\alpha\pi}{2}+k\omega^{\lambda-1}\sin\frac{\lambda\pi}{2}}{2\sqrt{mk\omega^{\alpha+\lambda-2}\left|\cos\frac{\alpha\pi}{2}\right|\cos\frac{\lambda\pi}{2}}}\sqrt{\frac{-\omega^{\alpha-2}\cos\frac{\alpha\pi}{2}}{\omega^{\lambda}\cos\frac{\lambda\pi}{2}}}\right)^2\right]}. \quad (1.24)$$

*Proof.* Because $S_{xx4}(\omega) = S_{ff}(\omega)|H_4(\omega)|^2$, where (Chapter 7 in Volume I or Li [1] or [2])

$$H_4(\omega)=\frac{1}{k\omega^{\lambda}\cos\frac{\lambda\pi}{2}\left(1-\gamma^2\frac{-\omega^{\alpha-2}\cos\frac{\alpha\pi}{2}}{\omega^{\lambda}\cos\frac{\lambda\pi}{2}}+i2\gamma\frac{m\omega^{\alpha-1}\sin\frac{\alpha\pi}{2}+k\omega^{\lambda-1}\sin\frac{\lambda\pi}{2}}{2\sqrt{mk\omega^{\alpha+\lambda-2}\left|\cos\frac{\alpha\pi}{2}\right|\cos\frac{\lambda\pi}{2}}}\sqrt{\frac{-\omega^{\alpha-2}\cos\frac{\alpha\pi}{2}}{\omega^{\lambda}\cos\frac{\lambda\pi}{2}}}\right)}, \quad (1.25)$$

(1.24) holds. This finishes the proof.

**Theorem 1.8 (cross-PSD response IV)**

Denote by $S_{fx4}(\omega)$ the cross PSD between $f(t)$ and $x_4(t)$. Then,

$$S_{fx4}(\omega)=\frac{ag^2\omega^{-5}e^{-b\left(\frac{g}{V}\right)^4\frac{1}{\omega^4}}}{\left(k\omega^{\lambda}\cos\frac{\lambda\pi}{2}\left[1-\gamma^2\frac{-\omega^{\alpha-2}\cos\frac{\alpha\pi}{2}}{\omega^{\lambda}\cos\frac{\lambda\pi}{2}}+i2\gamma\frac{m\omega^{\alpha-1}\sin\frac{\alpha\pi}{2}+k\omega^{\lambda-1}\sin\frac{\lambda\pi}{2}}{2\sqrt{mk\omega^{\alpha+\lambda-2}\left|\cos\frac{\alpha\pi}{2}\right|\left|\cos\frac{\lambda\pi}{2}\right|}}\sqrt{\frac{-\omega^{\alpha-2}\cos\frac{\alpha\pi}{2}}{\omega^{\lambda}\cos\frac{\lambda\pi}{2}}}\right]\right)}. \quad (1.26)$$

*Proof.* Doing the operation of $S_{fx4}(\omega)=S_{ff}(\omega)H_4(\omega)$ yields (1.26). The proof is finished.

Doing the inverse Fourier transform on both sides of $S_{xx4}(\omega) = S_{ff}(\omega)|H_4(\omega)|^2$ yields the ACF response in the form

$$r_{xx4}(\tau)=r_{ff}(\tau)*h_4(\tau)*h_4(-\tau). \quad (1.27)$$

In (1.27), $r_{xx4}(\tau)$ is the ACF of $x_4(t)$ and $h_4(\tau)$ is the impulse response of a class IV fractional vibrator. In addition, doing the inverse Fourier transform on both sides of $S_{fx4}(\omega)=S_{ff}(\omega)H_4(\omega)$ results in

$$r_{fx4}(\tau)=r_{ff}(\tau)*h_4(\tau). \quad (1.28)$$

In (1.28), $r_{fx4}(\tau)$ is the cross-correlation between $f(t)$ and $x_4(t)$. It is the cross-correlation response. In (1.27) and (1.28),

$$h_4(t)=e^{-\frac{m\omega^{\alpha-1}\sin\frac{\alpha\pi}{2}+k\omega^{\lambda-1}\sin\frac{\lambda\pi}{2}}{2\sqrt{mk\omega^{\alpha+\lambda-2}\left|\cos\frac{\alpha\pi}{2}\right|\left|\cos\frac{\lambda\pi}{2}\right|}}\sqrt{\frac{\omega^{\lambda}\cos\frac{\lambda\pi}{2}}{-\omega^{\alpha-2}\cos\frac{\alpha\pi}{2}}}\omega_n t}\frac{1}{m_{\mathrm{eq4}}\omega_{\mathrm{eqd4}}}\sin\omega_{\mathrm{eqd4}}tu(t). \quad (1.29)$$

Refer to Chapter 7 in Volume I or Li [1] or Li [2] for the expressions of $m_{eq4}$ and $\omega_{eqd4}$ in (1.29).

### 1.5.2 Effect of $(\alpha, \lambda)$ on Responses

We illustrate some plots of $|H_4(\omega)|^2$ in Figure 1.15. Figure 1.16 indicates some plots of $S_{xx4}(\omega)$. Some plots of $S_{fx4}(\omega)$ are shown in Figure 1.17. Figures 1.16 and 1.17 exhibit that the effect of fractional orders $(\alpha, \lambda)$ on the responses to class IV fractional vibrators driven by the P-M spectrum is noticeable.

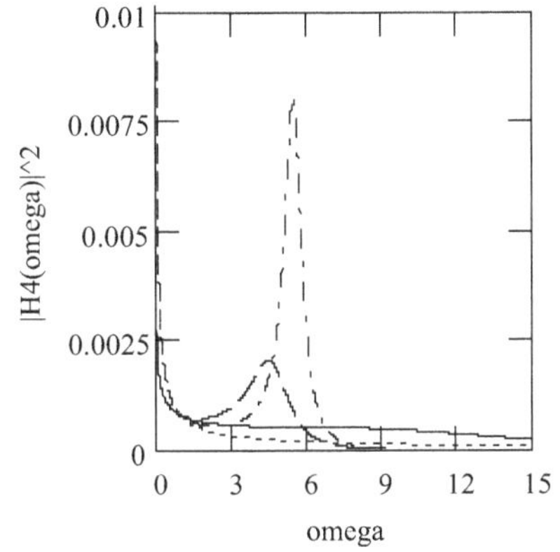


FIGURE 1.15 Plots of $|H_4(\omega)|^2$ when $m = 1$, $c = 0$, and $k = 36$, for $(\alpha, \lambda) = (1.6, 0.2)$ (solid), (1.6, 0.4) (dot), (2.5, 0.2) (dash), (2.5, 0.4) (dash dot).

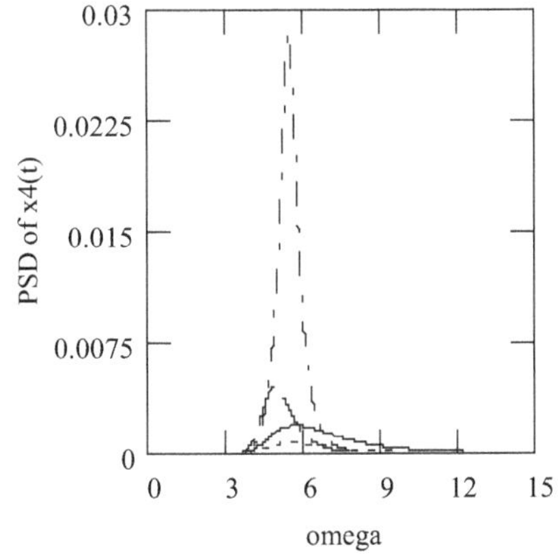


FIGURE 1.16 Plots of response PSD $S_{xx4}(\omega)$ when $m = 1$, $c = 0$, $k = 36$, and $V = 15$ for $(\alpha, \lambda) = (1.6, 0.2)$ (solid), (1.6, 0.4) (dot), (2.5, 0.2) (dash), (2.5, 0.4) (dash dot).

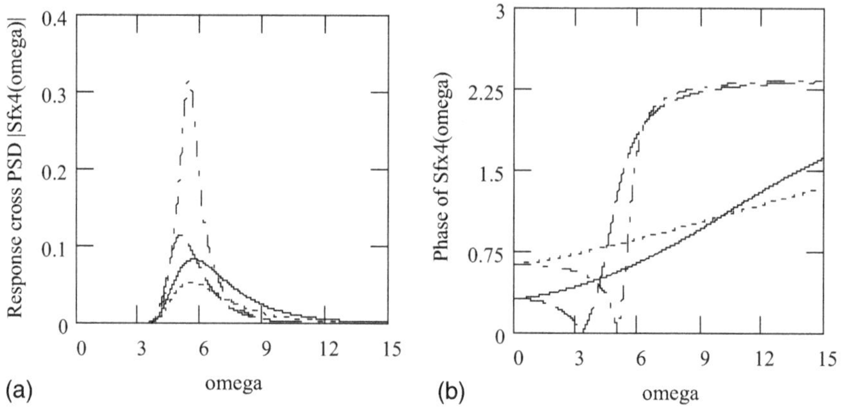


FIGURE 1.17 Plots of cross-PSD response $S_{fx4}(\omega)$ when $m = 1$, $c = 0$, $k = 36$, and $V = 15$ for $(\alpha, \lambda) = (1.6, 0.2)$ (solid), $(1.6, 0.4)$ (dot), $(2.5, 0.2)$ (dash), $(2.5, 0.4)$ (dash dot). (a). $|S_{fx4}(\omega)|$. (b). Phase of $S_{fx4}(\omega)$.

## 1.6 RESPONSES OF CLASS V FRACTIONAL VIBRATORS DRIVEN BY P-M SPECTRUM

### 1.6.1 Computations

Consider the following motion equation of a class V fractional vibrator

$$m\frac{d^2x_5(t)}{dt^2}+k\frac{d^\lambda x_5(t)}{dt^\lambda}=f(t). \tag{1.30}$$

In (1.30), where $x_5(t)$ is the response of a class V fractional vibrator.

**Theorem 1.9 (PSD response V)**

Denote by $S_{xx5}(\omega)$ the PSD of $x_5(t)$. Then,

$$S_{xx5}(\omega)=\frac{ag^2\omega^{-5}e^{-b\left(\frac{g}{V}\right)^4\frac{1}{\omega^4}}}{k^2\omega^{2\lambda}\cos^2\frac{\lambda\pi}{2}\left[\left(1-\frac{\gamma^2}{\omega^\lambda\cos\frac{\lambda\pi}{2}}\right)^2+4\gamma^2\left(\frac{k\omega^{\lambda-1}\sin\frac{\lambda\pi}{2}}{2\sqrt{mk\omega^\lambda\cos\frac{\lambda\pi}{2}}}\sqrt{\frac{1}{\omega^\lambda\cos\frac{\lambda\pi}{2}}}\right)^2\right]}. \tag{1.31}$$

*Proof.* Following Chapter 7 in Volume I or Li [1] or [2], we have

$$H_5(\omega)=\frac{1}{k\omega^{\lambda}\cos\frac{\lambda\pi}{2}\left(1-\frac{\frac{\gamma^2}{\omega^{\lambda}\cos\frac{\lambda\pi}{2}}}{\sqrt{\frac{1}{\omega^{\lambda}\cos\frac{\lambda\pi}{2}}}}+i2\gamma\frac{k\omega^{\lambda-1}\sin\frac{\lambda\pi}{2}}{2\sqrt{mk\omega^{\lambda}\cos\frac{\lambda\pi}{2}}}\right)}. \tag{1.32}$$

Taking into account (1.31) in $S_{xx5}(\omega) = S_{ff}(\omega)|H_5(\omega)|^2$ results in (1.32). The proof completes.

**Theorem 1.10 (cross-PSD response V)**

Let $S_{fx5}(\omega)$ be the cross PSD between $f(t)$ and $x_5(t)$. Then,

$$S_{fx5}(\omega)=\frac{ag^2\omega^{-5}e^{-b\left(\frac{g}{V}\right)^4\frac{1}{\omega^4}}}{k\omega^{\lambda}\cos\frac{\lambda\pi}{2}\left(1-\frac{\frac{\gamma^2}{\omega^{\lambda}\cos\frac{\lambda\pi}{2}}}{\sqrt{\frac{1}{\omega^{\lambda}\cos\frac{\lambda\pi}{2}}}}+i2\gamma\frac{k\omega^{\lambda-1}\sin\frac{\lambda\pi}{2}}{2\sqrt{mk\omega^{\lambda}\cos\frac{\lambda\pi}{2}}}\right)}. \tag{1.33}$$

*Proof.* Doing the operation of $S_{fx5}(\omega) = S_{ff}(\omega)H_5(\omega)$ yields (1.33). This finishes the proof.

Let $r_{xx5}(\tau)$ be the ACF of $x_5(t)$. Denote by $h_5(\tau)$ the impulse response of a class V fractional vibrator. Then, the ACF response $r_{xx5}(\tau)$ is given by

$$r_{xx5}(\tau) = r_{ff}(\tau)*h_5(\tau)*h_5(-\tau). \tag{1.34}$$

Let $r_{fx5}(\tau)$ be the cross-correlation between $f(t)$ and $x_5(t)$. Then, the cross-correlation response $r_{fx5}(\tau)$ is expressed by

$$r_{fx5}(\tau) = r_{ff}(\tau)*h_5(\tau). \tag{1.35}$$

In (1.34) and (1.35),

$$h_5(t) = e^{-\frac{k\omega^{\lambda-1}\sin\frac{\lambda\pi}{2}}{2\sqrt{mk\omega^{\lambda}\cos\frac{\lambda\pi}{2}}}\sqrt{\omega^{\lambda}\cos\frac{\lambda\pi}{2}}\omega_n t}\frac{1}{m\omega_{eqd5}}\sin\omega_{eqd5}tu(t). \qquad (1.36)$$

Refer to Chapter 7 in Volume I or Li [1] or [2] for the expression of $\omega_{eqd5}$ in (1.36).

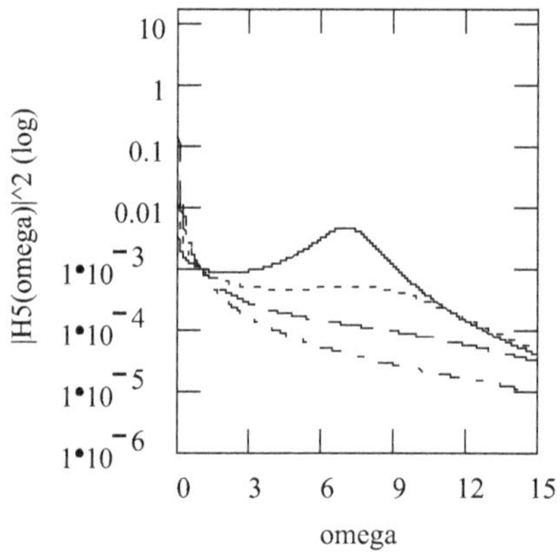


FIGURE 1.18 Plots of $|H_5(\omega)|^2$ (log) when $m = 1$, $c = 0$, and $k = 36$, for $\lambda = 0.2$ (solid), 0.4 (dot), 0.6 (dash), 0.8 (dash dot).

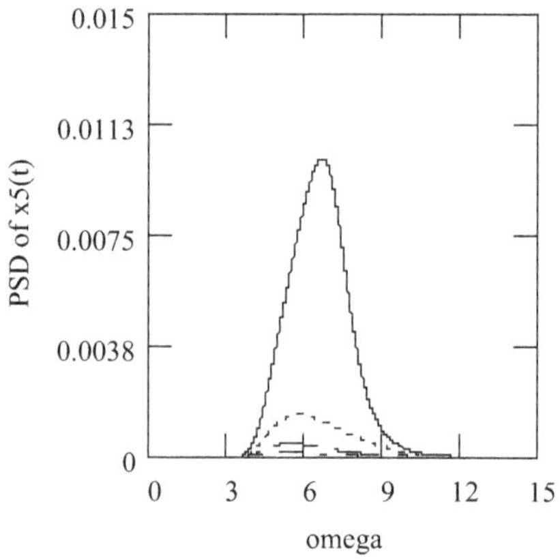


FIGURE 1.19 Plots of response PSD $S_{xx5}(\omega)$ when $m = 1$, $c = 0$, $k = 36$, and $V = 15$ for $\lambda = 0.2$ (solid), 0.4 (dot), 0.6 (dash), 0.8 (dash dot).

### 1.6.2 Effect of $\lambda$ on Responses

Figure 1.18 illustrates some plots of $|H_5(\omega)|^2$. Figure 1.19 shows some plots of $S_{xx5}(\omega)$ and Figure 1.20 indicates some plots of $S_{fx5}(\omega)$. Figures 1.19 and 1.20 show that the effect of the fractional order $\lambda$ on the responses to class V fractional vibration systems driven by the P-M spectrum is considerable.

## 1.7 RESPONSES OF CLASS VI FRACTIONAL VIBRATORS DRIVEN BY P-M SPECTRUM

### 1.7.1 Computations

The motion equation of a class VI fractional vibrator is expressed by

$$m\frac{d^{\alpha}x_6(t)}{dt^{\alpha}}+c\frac{d^{\beta}x_6(t)}{dt^{\beta}}+k\frac{d^{\lambda}x_6(t)}{dt^{\lambda}}=f(t). \tag{1.37}$$

In (1.37), $x_6(t)$ is the response of a class VI fractional vibrator.

**Theorem 1.11 (PSD response VI)**

Let $S_{xx6}(\omega)$ be the PSD of $x_6(t)$. Then,

$$S_{xx6}(\omega)=\frac{1}{k^2}\frac{ag^2\omega^{-5}e^{-b\left(\frac{g}{V}\right)^4\frac{1}{\omega^4}}}{\left[\omega^{\lambda}\cos\frac{\lambda\pi}{2}+\gamma^2\left(\omega^{\alpha-2}\cos\frac{\alpha\pi}{2}+2\varsigma\omega_n\omega^{\beta-2}\cos\frac{\beta\pi}{2}\right)\right]^2+\gamma^2\left(\omega^{\alpha-1}\sin\frac{\alpha\pi}{2}+2\zeta\omega_n\omega^{\beta-1}\sin\frac{\beta\pi}{2}+\omega_n^2\omega^{\lambda-1}\sin\frac{\lambda\pi}{2}\right)^2}. \tag{1.38}$$

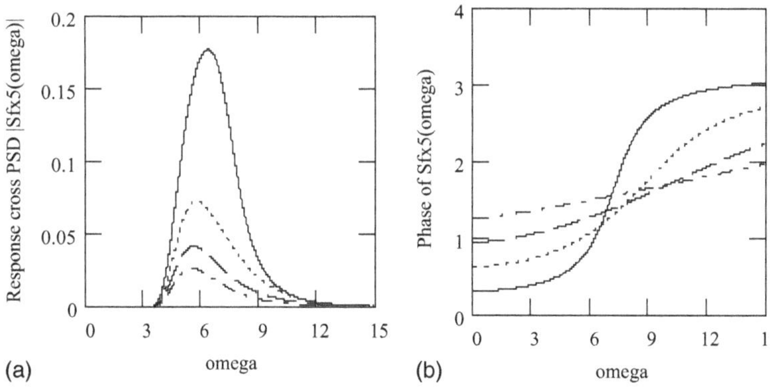


FIGURE 1.20 Plots of cross-PSD response $S_{fx5}(\omega)$ when $m = 1$, $c = 0$, $k = 36$, and $V = 15$ for $\lambda = 0.2$ (solid), 0.4 (dot), 0.6) (dash), 0.8 (dash dot). (a). $|S_{fx5}(\omega)|$. (b). Phase of $S_{fx5}(\omega)$.

*Proof.* Doing the operation of $S_{xx6}(\omega) = S_{ff}(\omega)|H_6(\omega)|^2$ produces (1.38), where $H_6(\omega)$ is given by

$$H_6(\omega)=\frac{1}{k\begin{bmatrix}\omega^{\lambda}\cos\frac{\lambda\pi}{2}+\gamma^2\left(\omega^{\alpha-2}\cos\frac{\alpha\pi}{2}+2\varsigma\omega_n\omega^{\beta-2}\cos\frac{\beta\pi}{2}\right)\\+i\gamma\left(\omega^{\alpha-1}\sin\frac{\alpha\pi}{2}+2\zeta\omega_n\omega^{\beta-1}\sin\frac{\beta\pi}{2}+\omega_n^2\omega^{\lambda-1}\sin\frac{\lambda\pi}{2}\right)\end{bmatrix}}. \quad (1.39)$$

The proof ends.

**Theorem 1.12 (cross-PSD response VI)**

Denote by $S_{fx6}(\omega)$ the cross-PSD between $f(t)$ and $x_6(t)$. Then,

$$S_{fx6}(\omega)=\frac{ag^2\omega^{-5}e^{-b\left(\frac{g}{V}\right)^4\frac{1}{\omega^4}}}{k\begin{bmatrix}\omega^{\lambda}\cos\frac{\lambda\pi}{2}+\gamma^2\left(\omega^{\alpha-2}\cos\frac{\alpha\pi}{2}+2\varsigma\omega_n\omega^{\beta-2}\cos\frac{\beta\pi}{2}\right)\\+i\gamma\left(\omega^{\alpha-1}\sin\frac{\alpha\pi}{2}+2\zeta\omega_n\omega^{\beta-1}\sin\frac{\beta\pi}{2}+\omega_n^2\omega^{\lambda-1}\sin\frac{\lambda\pi}{2}\right)\end{bmatrix}}. \quad (1.40)$$

*Proof.* Doing $S_{fx6}(\omega) = S_{ff}(\omega)H_6(\omega)$ results in (1.40). The proof ends.

Let $r_{xx6}(\tau)$ be the ACF of $x_6(t)$. Let $h_6(\tau)$ be the impulse response of a class VI fractional vibrator. Then, we have the ACF response $r_{xx6}(\tau)$ expressed by

$$r_{xx6}(\tau) = r_{ff}(\tau)*h_6(\tau)*h_6(-\tau). \quad (1.41)$$

Let $r_{fx6}(\tau)$ be the cross-correlation between $f(t)$ and $x_6(t)$. Then, the cross-correlation response $r_{fx6}(\tau)$ is given by

$$r_{fx6}(\tau) = r_{ff}(\tau)*h_6(\tau). \quad (1.42)$$

In (1.41) and (1.42),

$$h_6(t)=e^{-\varsigma_{\text{eq6}}\omega_{\text{eqn6}}t}\frac{1}{m_{\text{eq6}}\omega_{\text{eqd6}}}\sin\omega_{\text{eqd6}}tu(t). \quad (1.43)$$

Refer to Chapter 7 in Volume I or Li [1] or [2] for the expressions of $\zeta_{\text{eq6}}$, $\omega_{\text{eqn6}}$, and $\omega_{\text{eqd6}}$ in (1.43).

### 1.7.2 Effect of ($\alpha$, $\beta$, $\lambda$) on Responses

Figure 1.21 shows some plots of $|H_6(\omega)|^2$, Figure 1.22 gives some plots of response PSD $S_{xx6}(\omega)$, and Figure 1.23 indicates some plots of cross-PSD response $S_{fx6}(\omega)$. Figures 1.22 and 1.23 show that the effect of ($\alpha$, $\beta$, $\lambda$) on the responses of class VI fractional vibrators driven by the P-M spectrum is significant.

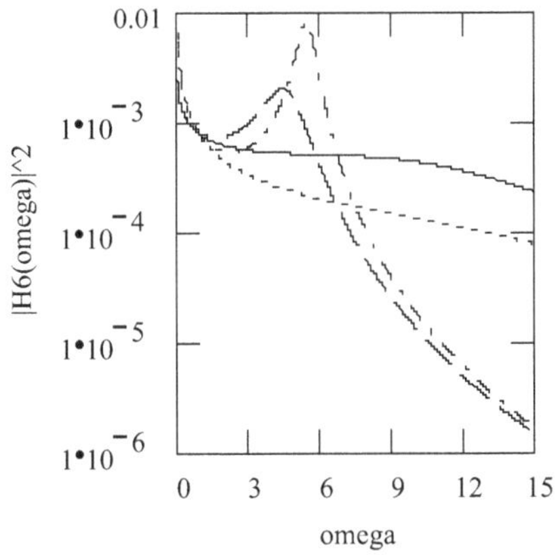


FIGURE 1.21 Plots of $|H_6(\omega)|^2$ (log) when $m = 1$, $c = 0.1$, and $k = 36$, for $(\alpha, \beta, \lambda) =$ (1.6, 0.8, 0.2) (solid), (1.6, 1.8, 0.4) (dot), (2.5, 0.4, 0.2) (dash), (2.5, 0.8, 0.4) (dash dot).

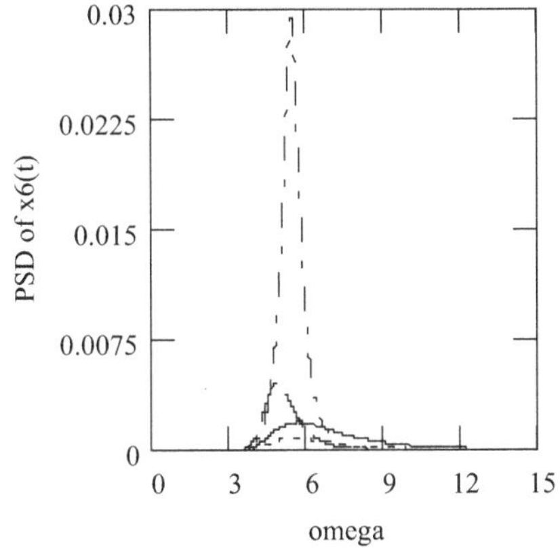


FIGURE 1.22 Plots of $S_{xx6}(\omega)$ when $m = 1$, $c = 0.1$, $k = 36$, and $V = 15$ for $(\alpha, \beta, \lambda) =$ (1.6, 0.8, 0.2) (solid), (1.6, 1.8, 0.4) (dot), (2.5, 0.4, 0.2) (dash), (2.5, 0.8, 0.4) (dash dot).

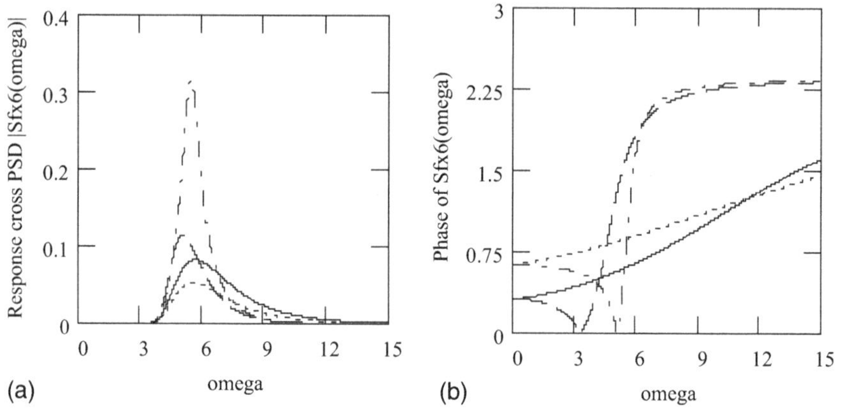


FIGURE 1.23 Plots of $S_{fx6}(\omega)$ when $m=1$, $c=0.1$, $k=36$, and $V=15$ for $(\alpha,\beta,\lambda)=$ (1.6, 0.8, 0.2) (solid), (1.6, 1.8, 0.4) (dot), (2.5, 0.4, 0.2) (dash), (2.5, 0.8, 0.4) (dash dot). (a). $|S_{fx6}(\omega)|$. (b). Phase of $S_{fx6}(\omega)$.

## 1.8 RESPONSES OF CLASS VII FRACTIONAL VIBRATORS DRIVEN BY P-M SPECTRUM

### 1.8.1 Computations

Consider the motion equation of a class VII fractional vibrator. It is expressed by

$$m\frac{d^2x_7(t)}{dt^2}+c\frac{d^\beta x_7(t)}{dt^\beta}+k\frac{d^\lambda x_7(t)}{dt^\lambda}=0. \tag{1.44}$$

In (1.44), $x_7(t)$ is the response of a class VII fractional vibrator.

**Theorem 1.13 (PSD response VII)**

Denote by $S_{xx7}(\omega)$ the PSD of $x_7(t)$. Then,

$$S_{xx7}(\omega)=\frac{1}{k^2}\frac{ag^2\omega^{-5}e^{-b\left(\frac{g}{V}\right)^4\frac{1}{\omega^4}}}{\left[\omega^\lambda\cos\frac{\lambda\pi}{2}-\gamma\left(1-2\varsigma\omega_n\omega^{\beta-2}\cos\frac{\beta\pi}{2}\right)\right]^2+\gamma^2\left(2\varsigma\omega^{\beta-1}\sin\frac{\beta\pi}{2}+\omega_n\omega^{\lambda-1}\sin\frac{\lambda\pi}{2}\right)^2}. \tag{1.45}$$

*Proof.* Performing $S_{xx7}(\omega) = S_{ff}(\omega)|H_7(\omega)|^2$ results in (1.45), where $H_7(\omega)$ is expressed by (Chapter 7 in Volume I or Li [2])

$$H_7(\omega) = \frac{1}{k\left[\begin{array}{l}\omega^{\lambda}\cos\frac{\lambda\pi}{2} - \gamma\left(1 - 2\varsigma\omega_n\omega^{\beta-2}\cos\frac{\beta\pi}{2}\right) \\ +i\gamma\left(2\varsigma\omega^{\beta-1}\sin\frac{\beta\pi}{2} + \omega_n\omega^{\lambda-1}\sin\frac{\lambda\pi}{2}\right)\end{array}\right]}. \tag{1.46}$$

This finishes the proof.

**Theorem 1.14 (cross-PSD response VII)**

Let $S_{fx7}(\omega)$ be the cross PSD between $f(t)$ and $x_7(t)$. Then,

$$S_{fx7}(\omega) = \frac{ag^2\omega^{-5}e^{-b\left(\frac{g}{V}\right)^4\frac{1}{\omega^4}}}{k\left[\begin{array}{l}\omega^{\lambda}\cos\frac{\lambda\pi}{2} - \gamma\left(1 - 2\varsigma\omega_n\omega^{\beta-2}\cos\frac{\beta\pi}{2}\right) \\ +i\gamma\left(2\varsigma\omega^{\beta-1}\sin\frac{\beta\pi}{2} + \omega_n\omega^{\lambda-1}\sin\frac{\lambda\pi}{2}\right)\end{array}\right]}. \tag{1.47}$$

*Proof.* Doing the operation of $S_{fx7}(\omega) = S_{ff}(\omega)H_7(\omega)$ yields (1.47). The proof ends.

Let $r_{xx7}(\tau)$ be the ACF of $x_7(t)$. Denote by $h_7(\tau)$ the impulse response of a class VII fractional vibrator. Then,

$$r_{xx7}(\tau) = r_{ff}(\tau) * h_7(\tau) * h_7(-\tau). \tag{1.48}$$

Let $r_{fx7}(\tau)$ be the cross-correlation between $f(t)$ and $x_7(t)$. Then,

$$r_{fx7}(\tau) = r_{ff}(\tau) * h_7(\tau). \tag{1.49}$$

In (1.48) and (1.49),

$$h_7(t) = e^{-\varsigma_{\text{eq}7}\omega_{\text{eqn}7}t}\frac{1}{m_{\text{eq}7}\omega_{\text{eqd}7}}\sin\omega_{\text{eqd}7}t, \quad t \geq 0. \tag{1.50}$$

Refer to Chapter 7 in Volume I or Li [2] for the expressions of $\zeta_{eq7}$, $\omega_{eqn7}$, and $\omega_{eqd7}$ in (1.50).

### 1.8.2 Effects of $\beta$, $\lambda$ on Responses

Figure 1.24 shows some plots of $|H_7(\omega)|^2$. Figure 1.25 indicates some plots of PSD response $S_{xx7}(\omega)$. Figure 1.26 demonstrates some plots of cross-PSD response $S_{fx7}(\omega)$. Figures 1.25 and 1.26 indicate that there are significant

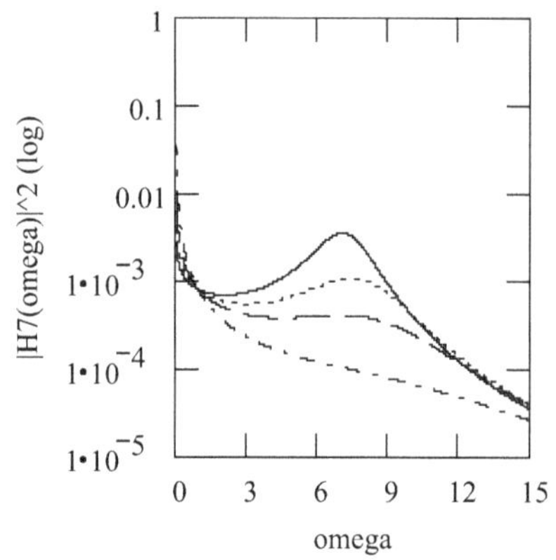


FIGURE 1.24 Plots of $|H_7(\omega)|^2$ (log) when $m = 1$, $c = 0.1$, and $k = 36$, for $(\beta, \lambda) =$ (0.5, 0.2) (solid), (0.5, 0.3) (dot), (1.5, 0.4) (dash), (1.5, 0.6) (dash dot).

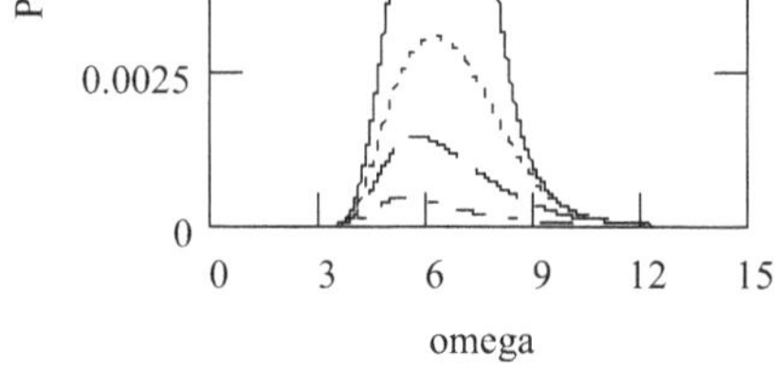

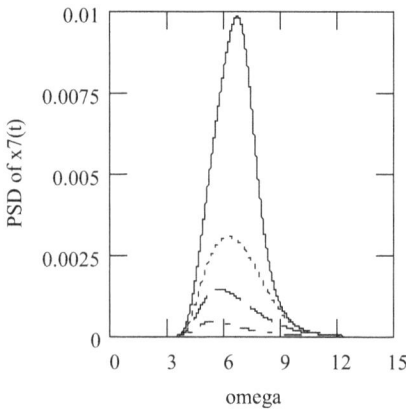

FIGURE 1.25 Plots of $S_{xx7}(\omega)$ when $m = 1$, $c = 0.1$, $k = 36$, and $V = 15$ for $(\beta, \lambda) =$ (0.5, 0.2) (solid), (0.5, 0.3) (dot), (1.5, 0.4) (dash), (1.5, 0.6) (dash dot).

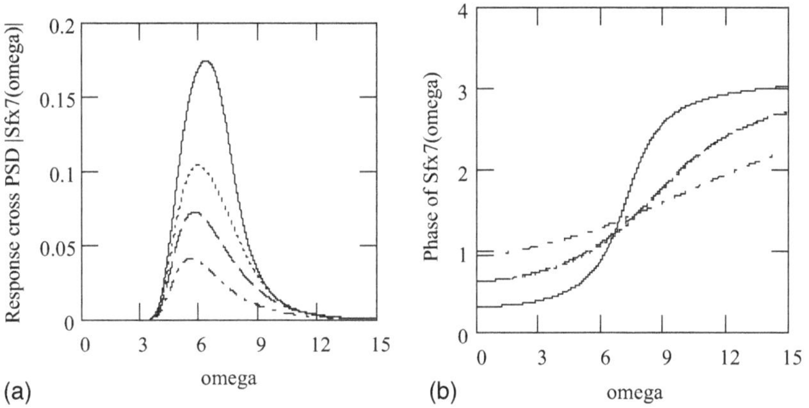


FIGURE 1.26 Plots of $S_{fx7}(\omega)$ when $m = 1$, $c = 0.1$, $k = 36$, and $V = 15$ for $(\beta, \lambda) =$ (0.5, 0.2) (solid), (0.5, 0.3) (dot), (1.5, 0.4) (dash), (1.5, 0.6) (dash dot). (a). $|S_{fx7}(\omega)|$. (b). Phase of $S_{fx7}(\omega)$.

effects of $(\beta, \lambda)$ on the responses to class VII fractional vibrators under the excitation of the P-M spectrum.

## 1.9 SUMMARY

We have presented the analytic expressions of the PSD and cross-PSD responses of fractional vibrators from class I to class VII in Theorems 1.1–1.14, respectively. We have shown that fractional orders of vibration systems, namely, $\alpha$, $\beta$, and $\lambda$, have considerable effects on responses.

## 1.10 EXERCISES

1.1. Suppose the excitation $f(t)$ follows the JONSWAP (Joint North Sea Wave Project) spectrum in the form

$$S_{ff}(\omega) = \alpha g^2 \omega^{-5} e^{-1.25\left(\frac{\omega_p}{\omega}\right)^4} \gamma^{\exp\left[\frac{(\omega-\omega_p)^2}{2u^2\omega^2}\right]},$$

where $\alpha = 8.1 \times 10^{-3}$, $g$ is the acceleration of gravity (m/s²), $F$ is fetch, and $V$ wind speed (m/s) at an elevation of 10 m above the sea surface, $1 < \gamma < 7$, $u = 0.07$ for $\omega \le \omega_p$, $u = 0.09$ for $\omega > \omega_p$, where $\omega_p = 2\pi \frac{g}{V}\left(\frac{V^2}{gF}\right)^{1/3}$. Find the inverse Fourier transform of $S_{ff}(\omega)$.

1.2. Let

$$h_1(t) = \frac{e^{-\frac{\omega \sin\frac{\alpha\pi}{2}}{2\left|\cos\frac{\alpha\pi}{2}\right|}t} \sin\left(\frac{\omega_n}{\sqrt{\omega^{\alpha-2}\left|\cos\frac{\alpha\pi}{2}\right|}}\sqrt{1-\frac{\omega^{2\alpha}\sin^2\frac{\alpha\pi}{2}}{4\omega_n^2\left|\cos\frac{\alpha\pi}{2}\right|}}t\right)}{m\omega_n\sqrt{\omega^{\alpha-2}\left|\cos\frac{\alpha\pi}{2}\right|}\sqrt{1-\frac{\omega^{2\alpha}\sin^2\frac{\alpha\pi}{2}}{4\omega_n^2\left|\cos\frac{\alpha\pi}{2}\right|}}} u(t),$$

where $u(t)$ is the unit step function. Find $h_1(\tau)*h_1(-\tau)$.

1.3. Denote by $r_{ff}(\tau)$ the inverse Fourier transform of $S_{ff}(\omega)$. Find $r_{xx1}(\tau) = r_{ff}(\tau)*h_1(\tau)*h_1(-\tau)$.

1.4. Find $r_{fx1}(\tau) = r_{ff}(\tau)*h_1(\tau)$.

## REFERENCES


1. M. Li, *Fractional Vibrations with Applications to Euler-Bernoulli Beams*, CRC Press, Boca Raton, 2023.
2. M. Li, Analytic theory of seven classes of fractional vibrations based on elementary functions: A tutorial review, *Symmetry*, 16(9): 2024, 1202.
3. J. J. Jensen, *Load and Global Response of Ships*, Vol. 4, Elsevier, Academic Press, Oxford, 2001.
4. Y. Okumoto, Y. Takeda, M. Mano, and T. Okada, *A Practical Guide for Engineers-Design of Ship Hull Structures*, Springer, Berlin, 2009.
5. W. J. Pierson and L. Moskowitz, A proposed spectral form for fully developed wind seas based on the similarity theory of S. A. Kitaigorodskii, *Journal of Geophysical Research*, 69(24): 1964, 5181–5190.
6. K. Hasselmann et al., Measurements of wind-wave growth and swell decay during the Joint North Sea Wave Project (JONSWAP), *Deutsche Hydrographische Zeitschrift*, 8(12): 1973, 95.
7. M. K. Ochi and E. N. Hubble, Six parameter wave spectra, *Proceedings of the 15th Coastal Engineering Conference*, Honolulu, HI, 1976, pp. 301–328.
8. J. R. Scott, A sea spectrum for model tests and long-term ship prediction, *Journal of Ship Research*, 9(2): 1965, 145–152.
9. O. M. Phillips, The equilibrium range in the spectrum of wind-generated waves, *Journal of Fluid Mechanics*, 4(4): 1958, 426–434.
10. S. R. Massel, *Ocean Surface Waves: Their Physics and Prediction*, World Scientific, Singapore, 1997.
11. S. K. Chairabarti, *Offshore Structure Modeling*, World Scientific, Singapore, 1994.

12. M. Li, A method for requiring block size for spectrum measurement of ocean surface waves, *IEEE Transactions on Instrumentation and Measurement*, 55(6): 2006, 2207–2215.
13. The Specialist Committee on Waves, Final report and recommendations to the 23rd ITTC, *Proceedings of the 23rd ITTC*, Vol. II, Venice, Italy, 2002, pp. 497–543.
14. M. Li, An iteration method to adjusting random loading for a laboratory fatigue test, *International Journal of Fatigue*, 27(7): 2005, 783–789.
15. M. Li, An optimal controller of an irregular wave maker, *Applied Mathematical Modelling*, 29(1): 2005, 55–63.
16. M. Li, *Multi-Fractal Traffic and Anomaly Detection in Computer Communications*, CRC Press, Boca Raton, 2022.
17. M. Li, Generation of teletraffic of generalized Cauchy type, *Physica Scripta*, 81(2): 2010, 025007, 10.
18. M. Li, PSD and cross PSD of responses of seven classes of fractional vibrations driven by fGn, fBm, fractional OU process, and von Kármán process, *Symmetry*, 16(5): 2024, 635.

CHAPTER 2

# Responses of Fractional Vibrations Driven by Fractional Gaussian Noise

THIS CHAPTER GIVES THE contributions threefold. First, we present the analytic expressions of the power spectrum density (PSD) and cross-PSD responses to seven classes of fractional vibration systems driven by fractional Gaussian noise (fGn). Second, we show that the orders of fractional vibrators have considerable effects on the responses. Last, we show that the statistical dependences of responses follow those of fGn.

## 2.1 BACKGROUND

Chapter 7 in Volume I discusses the results of seven classes of fractional vibration systems in closed-form expressions using elementary functions based on the theory stated in Li [1] and [2]. Though the literature of systems driven by fractional Gaussian noise (fGn) is quite rich, see, for example, Sun et al. [3, 4], Wang et al. [5], Hu and Zhou [6], and Liu et al. [7], closed-form expressions of PSD and cross-PSD responses to the excitation of fGn regarding seven classes of fractional vibration systems are seldom seen. This chapter gives the closed-form expressions of PSD and cross-PSD responses to seven classes of fractional vibration systems driven by fGn.


DOI: 10.1201/9781003657903-2

The rest of the chapter is organized as follows. In Sections 2.2–2.8, we present the closed-form expressions of PSD and cross-PSD responses to fractional vibration systems from classes I to VII under the excitation of fGn. The summary is given in Section 2.9.

## 2.2 RESPONSES OF CLASS I FRACTIONAL VIBRATORS DRIVEN BY FGN

### 2.2.1 Computations

Consider the motion equation of a class I fractional vibrator in the form

$$m\frac{d^{\alpha}x_1(t)}{dt^{\alpha}}+k\frac{dx_1(t)}{dt}=f(t). \tag{2.1}$$

In (2.1), $1 < \alpha < 3$, $x_1(t)$ is the response of a class I fractional vibrator, $f(t)$ is the driven force signal, $m$ and $k$ are the primary mass and stiffness, respectively.

Let $f(t)$ be fGn in what follows. Let $S_{ff}(\omega)$ be the PSD of $f(t)$. Then,

$$S_{ff}(\omega)=V_H\sin(H\pi)\Gamma(2H+1)|\omega|^{1-2H}. \tag{2.2}$$

In (2.2), $H\in(0,1)$ is the Hurst parameter, $V_H=\Gamma(1-2H)\dfrac{\cos\pi H}{\pi H}$ is the strength of fGn (Li [8–10], Li and Lim [11]). Figure 2.1 indicates some plots of $S_{ff}(\omega)$.

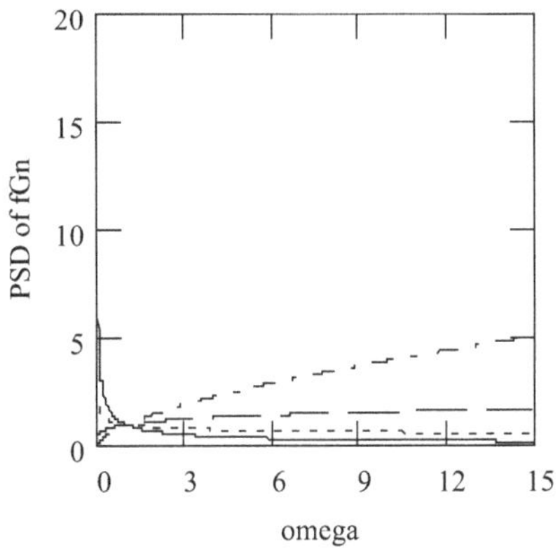


FIGURE 2.1 Plots of PSD of fGn for $H = 0.8$ (solid), 0.6 (dot), 0.4 (dash), 0.2 (dash dot).

**Theorem 2.1 (PSD response I)**

Denote by $S_{xx1}(\omega)$ the PSD of $x_1(t)$. Then,

$$S_{xx1}(\omega)=\frac{V_H\sin(H\pi)\Gamma(2H+1)|\omega|^{1-2H}}{k^2\left[\left(1-\frac{\omega^\alpha}{\omega_n^2}\left|\cos\frac{\alpha\pi}{2}\right|\right)^2+\left(\frac{\omega^\alpha}{\omega_n^2}\sin\frac{\alpha\pi}{2}\right)^2\right]}, \tag{2.3}$$

where $\omega_n^2=\frac{k}{m}$.

*Proof.* Note that $S_{xx1}(\omega)=S_{ff}(\omega)|H_1(\omega)|^2$. Following Chapter 7 in Volume I or Li [1] or [2], $H_1(\omega)$ is given

$$H_1(\omega)=\frac{1}{k\left(1-\frac{\omega^\alpha}{\omega_n^2}\left|\cos\frac{\alpha\pi}{2}\right|+i\frac{\omega^\alpha}{\omega_n^2}\sin\frac{\alpha\pi}{2}\right)}. \tag{2.4}$$

Thus, (2.3) results. The proof is finished.

**Theorem 2.2 (cross-PSD response I)**

Let $S_{fx1}(\omega)$ be the cross PSD between $f(t)$ and $x_1(t)$. Then,

$$S_{fx1}(\omega)=\frac{V_H\sin(H\pi)\Gamma(2H+1)|\omega|^{1-2H}}{k\left(1-\frac{\omega^\alpha}{\omega_n^2}\left|\cos\frac{\alpha\pi}{2}\right|+i\frac{\omega^\alpha}{\omega_n^2}\sin\frac{\alpha\pi}{2}\right)}. \tag{2.5}$$

*Proof.* Eq. (2.5) holds since $S_{fx1}(\omega)=S_{ff}(\omega)H_1(\omega)$. This finishes the proof.

Let $r_{ff}(\tau)$ be the autocorrelation function (ACF) of $f(t)$. Denote by $r_{xx1}(\tau)$ the ACF of $x_1(t)$. Let $h_1(\tau)$ be the impulse response of a class I fractional vibrator. Then, the ACF response $r_{xx1}(\tau)$ is given by

$$r_{xx1}(\tau)=r_{ff}(\tau)*h_1(\tau)*h_1(-\tau). \tag{2.6}$$

In (2.6), $*$ designates the convolution operation. The cross-correlation response, denoted by $r_{fx1}(\tau)$, is

$$r_{fx1}(\tau)=r_{ff}(\tau)*h_1(\tau). \tag{2.7}$$

In (2.6) and (2.7), $h_1(\tau)$ is actually the inverse Fourier transform of $H_1(\omega)$. Following Chapter 7 in Volume I or Li [1] or [2],

$$h_1(t) = \frac{e^{-\frac{\omega \sin\frac{\alpha\pi}{2}}{2\left|\cos\frac{\alpha\pi}{2}\right|}t} \sin\left(\frac{\omega_n}{\sqrt{\omega^{\alpha-2}\left|\cos\frac{\alpha\pi}{2}\right|}}\sqrt{1-\frac{\omega^{2\alpha}\sin^2\frac{\alpha\pi}{2}}{4\omega_n^2\left|\cos\frac{\alpha\pi}{2}\right|}}t\right)}{m\omega_n\sqrt{\omega^{\alpha-2}\left|\cos\frac{\alpha\pi}{2}\right|}\sqrt{1-\frac{\omega^{2\alpha}\sin^2\frac{\alpha\pi}{2}}{4\omega_n^2\left|\cos\frac{\alpha\pi}{2}\right|}}} u(t). \quad (2.8)$$

In (2.8), $u(t)$ is the unit step function.

### 2.2.2 Effect of $\alpha$ on Responses

Figure 2.2 indicates some plots of $|H_1(\omega)|^2$. Some plots of $S_{xx1}(\omega)$ are shown in Figure 2.3.

When $\alpha = 2$, a class I fractional vibrator reduces to be a conventional damping-free vibrator. Note that $H$ has a considerable effect on the responses, as can be seen from Figure 2.4. In Figure 2.5, we illustrate some plots of $S_{fx1}(\omega)$.

According to the generation method of long-range dependent (LRD) series introduced by Li [12], Li and Chi [13], we illustrate some plots of driven fGn and response ones in Figure 2.6. From Figures 2.3, 2.5, and 2.6, we see that there is a noticeable effect of the fractional order $\alpha$ on the responses of class I fractional vibration systems driven by fGn.

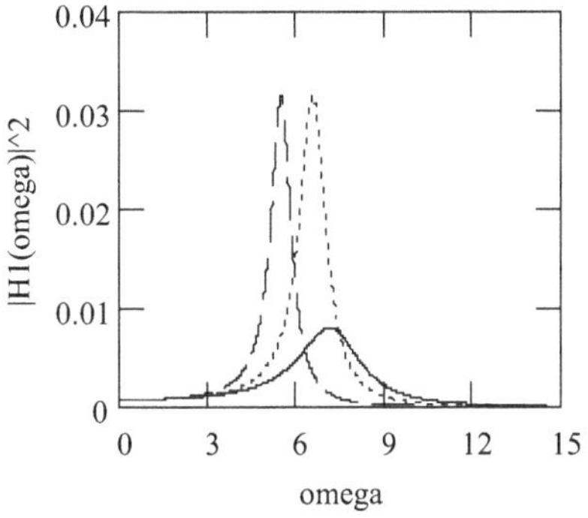


FIGURE 2.2 Plots of $|H_1(\omega)|^2$ with $\alpha = 1.8$ (solid), 1.9 (dot), 2.1 (dash) when $m = 1$ and $k = 36$.

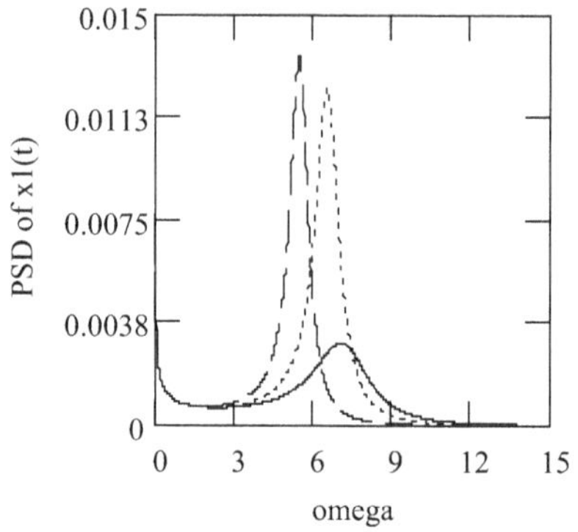


FIGURE 2.3 PSD response $S_{xx1}(\omega)$ with $\alpha = 1.8$ (solid), 1.9 (dot), 2.1 (dash) when $m = 1$, $k = 36$, and $H = 0.75$.

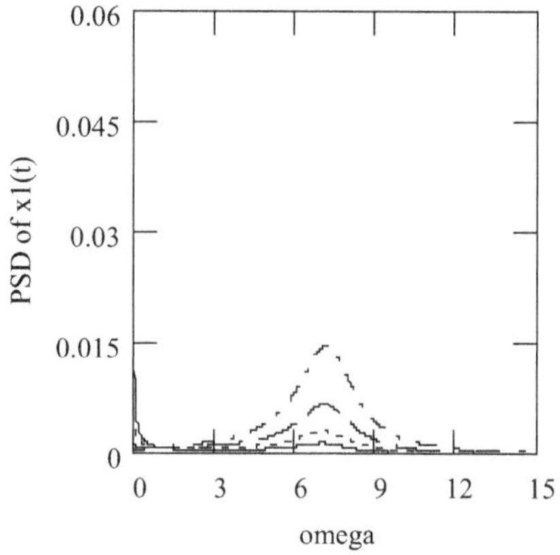


FIGURE 2.4 Observing $H$ effect on response PSD $S_{xx1}(\omega)$ when $\alpha = 1.8$, $m = 1$, and $k = 36$ for $H = 0.95$ (solid), 0.75 (dot), 0.55 (dash), 0.35 (dash dot).

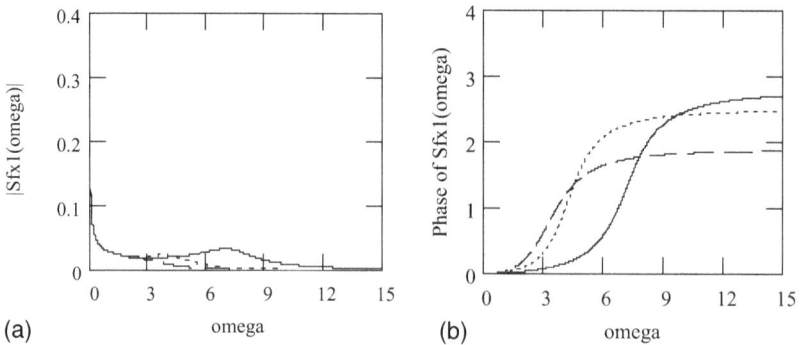

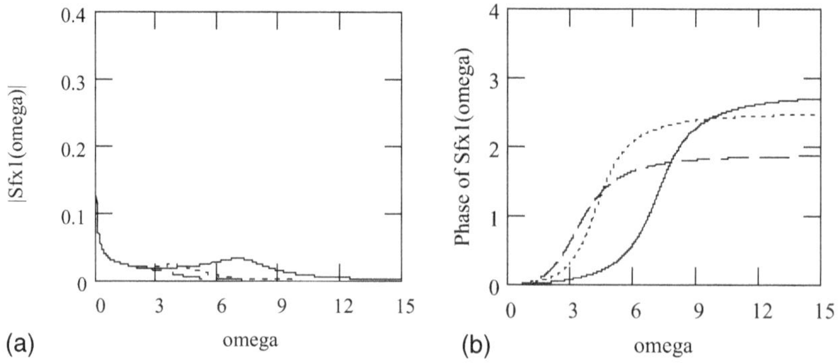


FIGURE 2.5 Cross-PSD response $S_{fx1}(\omega)$ with $\alpha = 1.8$ (solid), 2.4 (dot), 2.8 (dash) when $m = 1$, $k = 36$, and $H = 0.75$. (a). $|S_{fx1}(\omega)|$. (b). Phase of $S_{fx1}(\omega)$.

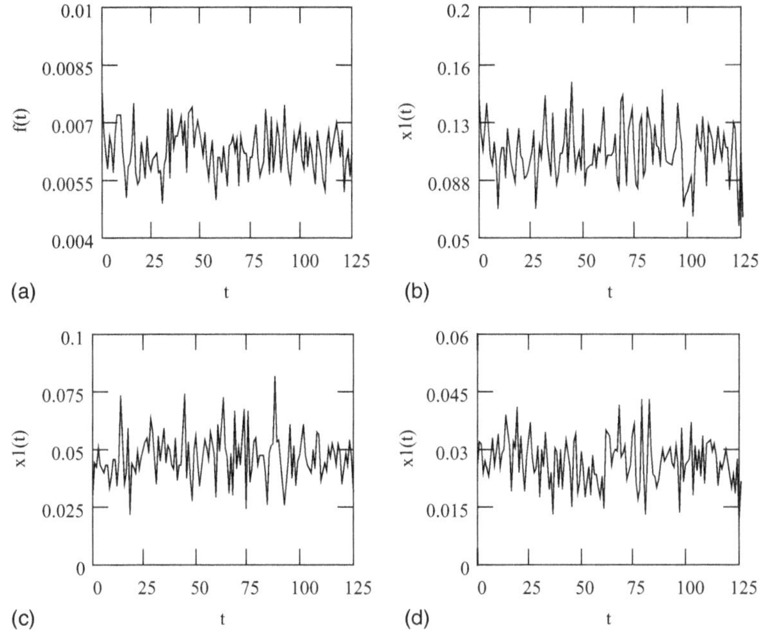


FIGURE 2.6 Plots of driven fGn and response signals when $m = 1$, $k = 36$, and $H = 0.75$. (a). Driven fGn. (b). Response $x_1(t)$ for $\alpha = 1.8$. (c). Response $x_1(t)$ for $\alpha = 2.4$. (d). Response $x_1(t)$ for $\alpha = 2.8$.

## 2.3 RESPONSES OF CLASS II FRACTIONAL VIBRATORS DRIVEN BY FGN

### 2.3.1 Computation Methods

For a class II fractional vibrator, its motion equation is given by

$$m\frac{d^2x_2(t)}{dt^2}+c\frac{d^\beta x_2(t)}{dt^\beta}+k\frac{dx_2(t)}{dt}=f(t). \tag{2.9}$$

In (2.9), $0 < \beta < 2$, $x_2(t)$ is the response of a class II fractional vibrator and $c$ is the primary damping.

**Theorem 2.3 (PSD response II)**

Let $S_{xx2}(\omega)$ be the PSD of $x_2(t)$. Then,

$$S_{xx2}(\omega)=\frac{V_H\sin(H\pi)\Gamma(2H+1)|\omega|^{1-2H}}{k^2\left\{\left[1-\gamma^2\left(1-\frac{c}{m}\omega^{\beta-2}\cos\frac{\beta\pi}{2}\right)\right]^2+\left(\frac{2\varsigma\omega^\beta}{\omega_n}\sin\frac{\beta\pi}{2}\right)^2\right\}}, \tag{2.10}$$

where $\gamma = \dfrac{\omega}{\omega_n}$.

*Proof.* Consider $S_{xx2}(\omega) = S_{ff}(\omega)|H_2(\omega)|^2$, where $H_2(\omega)$ (Chapter 7 in Volume I or Li [1] or [2])

$$H_2(\omega) = \frac{1/k}{1-\gamma^2\left(1-\dfrac{c}{m}\omega^{\beta-2}\cos\dfrac{\beta\pi}{2}\right)+i\dfrac{2\varsigma\omega^{\beta}\sin\dfrac{\beta\pi}{2}}{\omega_n}}. \tag{2.11}$$

Thus, (2.10) holds. The proof completes.

**Theorem 2.4 (cross-PSD response II)**

Denote by $S_{fx2}(\omega)$ the cross PSD between $f(t)$ and $x_2(t)$. Then,

$$S_{fx2}(\omega) = \frac{V_H \sin(H\pi)\Gamma(2H+1)|\omega|^{1-2H}}{k\left[1-\gamma^2\left(1-2\varsigma\omega_n\omega^{\beta-2}\cos\dfrac{\beta\pi}{2}\right)+i\dfrac{2\varsigma\omega^{\beta}}{\omega_n}\sin\dfrac{\beta\pi}{2}\right]}. \tag{2.12}$$

*Proof.* Doing the operation of $S_{fx2}(\omega) = S_{ff}(\omega)H_2(\omega)$ yields (2.12). The proof ends.

Denote by $r_{xx2}(\tau)$ the ACF response. Let $h_2(\tau)$ be the impulse response of a class II fractional vibrator. Then,

$$r_{xx2}(\tau) = r_{ff}(\tau) * h_2(\tau) * h_2(-\tau). \tag{2.13}$$

The cross-correlation response, denoted by $r_{fx2}(\tau)$, is

$$r_{fx2}(\tau) = r_{ff}(\tau) * h_2(\tau). \tag{2.14}$$

In (2.13) and (2.14), according to Chapter 7 in Volume I or Li [1] or [2],

$$h_2(t) = \frac{e^{-\frac{\varsigma\omega_n\omega^{\beta-1}\sin\frac{\beta\pi}{2}}{1-\frac{c}{m}\omega^{\beta-2}\cos\frac{\beta\pi}{2}}t}\sin\left(\dfrac{\omega_n\sqrt{1-\dfrac{\varsigma^2\omega^{2(\beta-1)}\sin^2\dfrac{\beta\pi}{2}}{1-\dfrac{c}{m}\omega^{\beta-2}\cos\dfrac{\beta\pi}{2}}}}{\sqrt{1-\dfrac{c}{m}\omega^{\beta-2}\cos\dfrac{\beta\pi}{2}}}t\right)}{\omega_n m\sqrt{1-\dfrac{c}{m}\omega^{\beta-2}\cos\dfrac{\beta\pi}{2}}\sqrt{1-\dfrac{\varsigma^2\omega^{2(\beta-1)}\sin^2\dfrac{\beta\pi}{2}}{1-\dfrac{c}{m}\omega^{\beta-2}\cos\dfrac{\beta\pi}{2}}}}u(t). \tag{2.15}$$

### 2.3.2 Effect of $\beta$ on Responses

Some plots of $|H_2(\omega)|^2$ are indicated in Figure 2.7. Figure 2.8 shows some plots of PSD response $S_{xx2}(\omega)$. Figure 2.9 shows a resonance curve when $\beta = 1$. Figure 2.10 illustrates some plots of cross-PSD response $S_{fx2}(\omega)$. Figure 2.11 exhibits the effects of $\beta$ on the fluctuation range of the response $x_2(t)$. Figures 2.8, 2.10, and 2.11 exhibit that there is effect of the fractional order $\beta$ on the responses of class II fractional vibration systems driven by fGn.

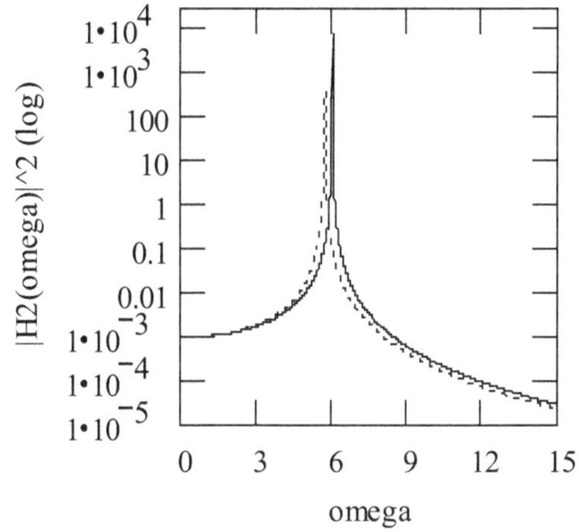


FIGURE 2.7 Plots of $|H_2(\omega)|^2$ (log) with $\beta = 0.01$ (solid), 1.99 (dot), when $m = 1$, $c = 0.1$, and $k = 36$.

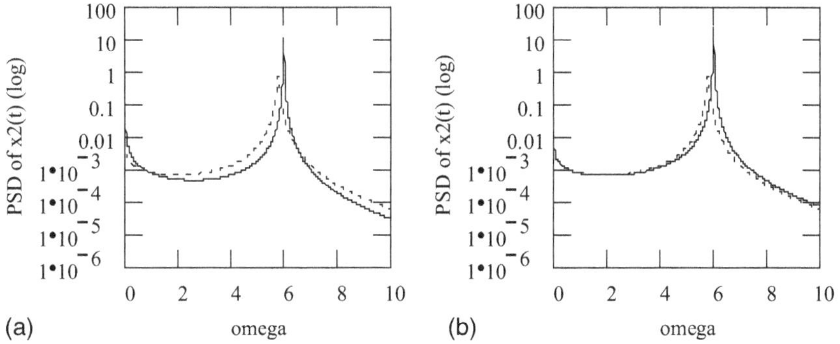


FIGURE 2.8 Response PSD $S_{xx2}(\omega)$ (log) with $\beta = 0.4$ (solid), 1.8 (dot), when $m = 1$, $c = 0.1$, and $k = 36$. (a). $S_{xx2}(\omega)$ when $H = 0.95$. (b). $S_{xx2}(\omega)$ when $H = 0.75$. (c). $S_{xx2}(\omega)$ when $H = 0.55$. (d). $S_{xx2}(\omega)$ when $H = 0.35$.

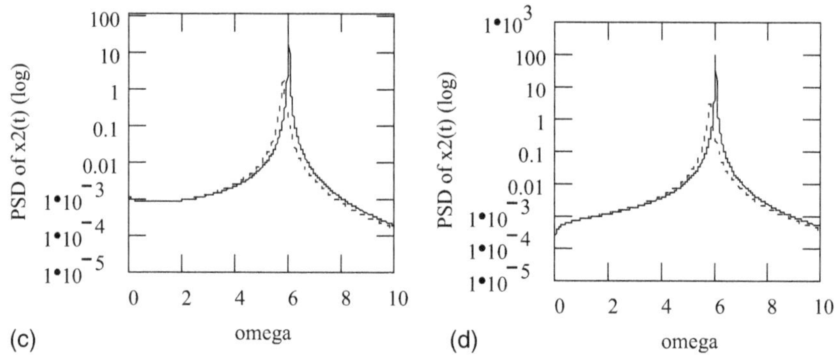


FIGURE 2.8 (Continued)

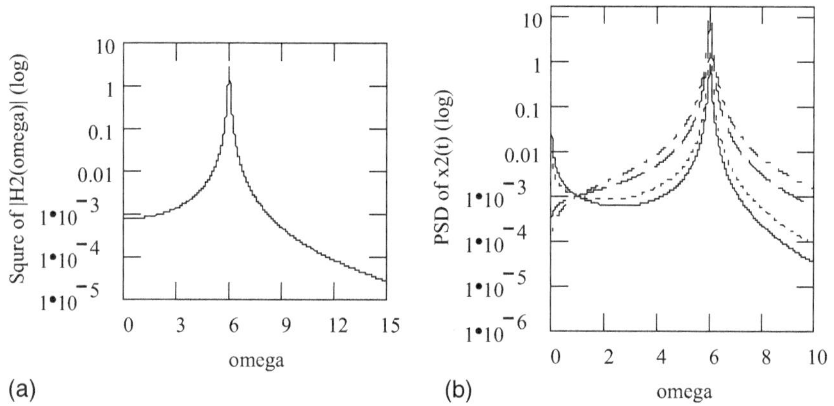


FIGURE 2.9 Resonance when with $\beta = 1$, $m = 1$, $c = 0.1$, $k = 36$, and $V = 15$. (a). Plot of $|H_2(\omega)|^2$ (log). (b). Response PSD $S_{xx2}(\omega)$ (log) for $H = 0.95$ (solid), 0.75 (dot), 0.35 (dash), 0.15 (dash dot).

## 2.4 RESPONSES OF CLASS III FRACTIONAL VIBRATORS DRIVEN BY FGN

### 2.4.1 Computations

The motion equation of a class III fractional vibrator is given by

$$m\frac{d^{\alpha}x_3(t)}{dt^{\alpha}} + c\frac{d^{\beta}x_3(t)}{dt^{\beta}} + kx_3(t) = f(t). \tag{2.16}$$

In (2.16), $x_3(t)$ is the response of a class III fractional vibrator.

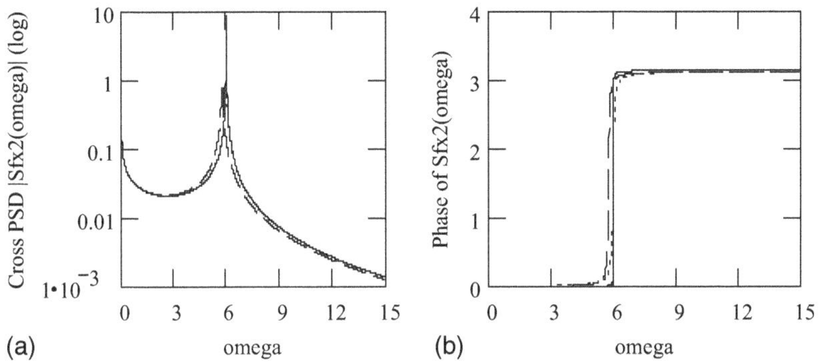


FIGURE 2.10 Cross-PSD response $S_{fx2}(\omega)$ with $\beta = 0.2$ (solid), 1 (dot), 1.9 (dash), when $m = 1$, $c = 0.1$, $k = 36$, and $H = 0.75$. (a). $|S_{fx2}(\omega)|$ (log). (b). Phase of $S_{fx2}(\omega)$.

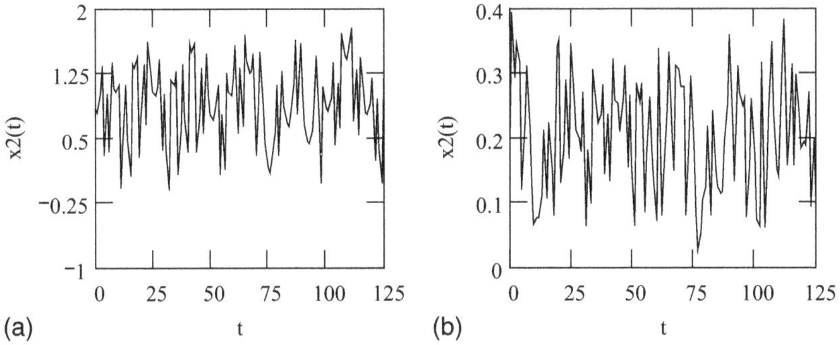


FIGURE 2.11 Response $x_2(t)$ when $m = 1$, $c = 0.1$, $k = 36$, and $H = 0.75$. (a). $x_2(t)$ for $\beta = 0.4$. (b). $x_2(t)$ for $\beta = 1.4$.

**Theorem 2.5 (PSD response III)**

Let $S_{xx3}(\omega)$ be the PSD of $x_3(t)$. Then,

$$S_{xx3}(\omega) = \frac{V_H \sin(H\pi)\Gamma(2H+1)|\omega|^{1-2H}}{k^2 \left\{ \left[1 - \gamma^2 \left(\omega^{\alpha-2}\left|\cos\frac{\alpha\pi}{2}\right| - 2\varsigma\omega_n\omega^{\beta-2}\cos\frac{\beta\pi}{2}\right)\right]^2 + \left[\frac{\gamma\left(\omega^{\alpha-1}\sin\frac{\alpha\pi}{2} + 2\varsigma\omega_n\omega^{\beta-1}\sin\frac{\beta\pi}{2}\right)}{\omega_n\left(\omega^{\alpha-2}\left|\cos\frac{\alpha\pi}{2}\right| - 2\varsigma\omega_n\omega^{\beta-2}\cos\frac{\beta\pi}{2}\right)}\right]^2 \right\}}. \quad (2.17)$$

*Proof.* Considering $S_{xx3}(\omega) = S_{ff}(\omega)|H_3(\omega)|^2$, where $H_3(\omega)$ is given by (Chapter 7 in Volume I or Li [1] or [2])

$$H_3(\omega) = \frac{1/k}{1-\gamma^2\left(\omega^{\alpha-2}\left|\cos\frac{\alpha\pi}{2}\right| - 2\varsigma\omega_n\omega^{\beta-2}\cos\frac{\beta\pi}{2}\right) + i\frac{\gamma\left(\omega^{\alpha-1}\sin\frac{\alpha\pi}{2} + 2\varsigma\omega_n\omega^{\beta-1}\sin\frac{\beta\pi}{2}\right)}{\omega_n\left(\omega^{\alpha-2}\left|\cos\frac{\alpha\pi}{2}\right| - 2\varsigma\omega_n\omega^{\beta-2}\cos\frac{\beta\pi}{2}\right)}}, \tag{2.18}$$

(2.17) results. The proof is finished.

**Theorem 2.6 (cross-PSD response III)**

Let $S_{fx3}(\omega)$ be the cross PSD between $f(t)$ and $x_3(t)$. Then,

$$S_{fx3}(\omega) = \frac{k^{-1}V_H\sin(H\pi)\Gamma(2H+1)|\omega|^{1-2H}}{1-\gamma^2\left(\omega^{\alpha-2}\left|\cos\frac{\alpha\pi}{2}\right| - 2\varsigma\omega_n\omega^{\beta-2}\cos\frac{\beta\pi}{2}\right) + i\frac{\gamma\left(\omega^{\alpha-1}\sin\frac{\alpha\pi}{2} + 2\varsigma\omega_n\omega^{\beta-1}\sin\frac{\beta\pi}{2}\right)}{\omega_n\left(\omega^{\alpha-2}\left|\cos\frac{\alpha\pi}{2}\right| - 2\varsigma\omega_n\omega^{\beta-2}\cos\frac{\beta\pi}{2}\right)}}. \tag{2.19}$$

*Proof.* Performing the operation of $S_{fx3}(\omega) = S_{ff}(\omega)H_3(\omega)$ and considering (2.18) produces (2.19) (Li [14]). The proof completes.

Let $r_{xx3}(\tau)$ be the ACF response in terms of $x_3(t)$. It is expressed by

$$r_{xx3}(\tau) = r_{ff}(\tau) * h_3(\tau) * h_3(-\tau). \tag{2.20}$$

In (2.20), $h_3(\tau)$ is the impulse response of a class III fractional vibrator. Denote by $r_{fx3}(\tau)$ the cross-correlation between $f(t)$ and $x_3(t)$. It is the cross-correlation response in the form

$$r_{fx3}(\tau) = r_{ff}(\tau) * h_3(\tau). \tag{2.21}$$

In (2.20) and (2.21), following Chapter 7 in Volume I or Li [1] or [2],

$$h_3(t) = \frac{e^{-\frac{m\omega^{\alpha-1}\sin\frac{\alpha\pi}{2}+c\omega^{\beta-1}\sin\frac{\beta\pi}{2}}{2\sqrt{-\left(m\omega^{\alpha-2}\cos\frac{\alpha\pi}{2}+c\omega^{\beta-2}\cos\frac{\beta\pi}{2}\right)k}}\omega_{\mathrm{eqn3}}t}\sin\omega_{\mathrm{eqd3}}t}{-\left(m\omega^{\alpha-2}\cos\frac{\alpha\pi}{2}+c\omega^{\beta-2}\cos\frac{\beta\pi}{2}\right)\omega_{\mathrm{eqd3}}}u(t). \tag{2.22}$$

Refer to Chapter 7 in Volume I or Li [1] or [2] for the expressions of $\omega_{\mathrm{eqn3}}$ and $\omega_{\mathrm{eqd3}}$ in (2.22).

### 2.4.2 Effect of ($\alpha$, $\beta$) on Responses

Some plots of $|H_3(\omega)|^2$ are shown in Figure 2.12. Figure 2.13 indicates some plots of $S_{xx3}(\omega)$. Some plots of $S_{fx3}(\omega)$ are indicated in Figure 2.14. Figures 2.13 and 2.14 show that there is significant effect of ($\alpha$, $\beta$) on the responses of a class III fractional vibrator driven by fGn.

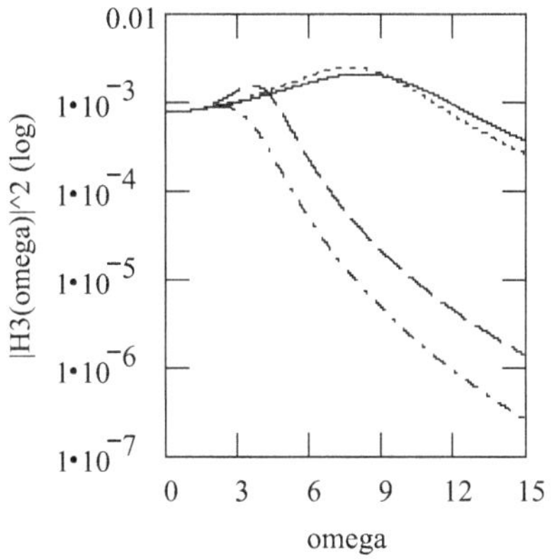


FIGURE 2.12 Plots of $|H_3(\omega)|^2$ in log when $m = 1$, $c = 0.1$, and $k = 36$, for $(\alpha, \beta) =$ (1.6, 0.8) (solid), (1.6, 1.8) (dot), (2.5, 0.8) (dash), (2.8, 1.8) (dash dot).

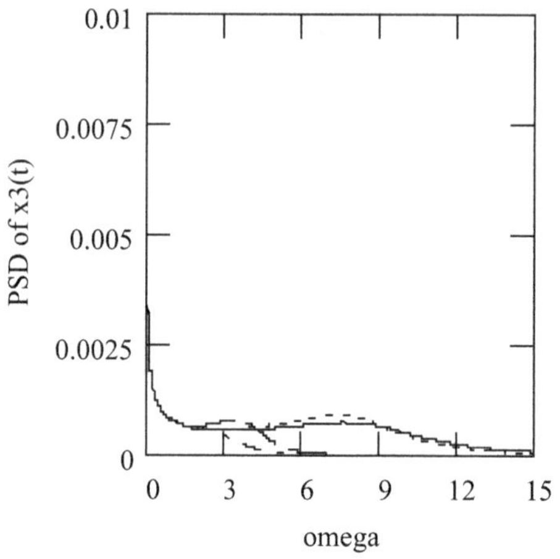


FIGURE 2.13 Response PSD $S_{xx3}(\omega)$ when $m=1$, $c=0.1$, $k=36$, and $H=0.75$ for $(\alpha, \beta)=(1.6, 0.8)$ (solid), (1.6, 1.8) (dot), (2.5, 0.8) (dash), (2.8, 1.8) (dash dot).

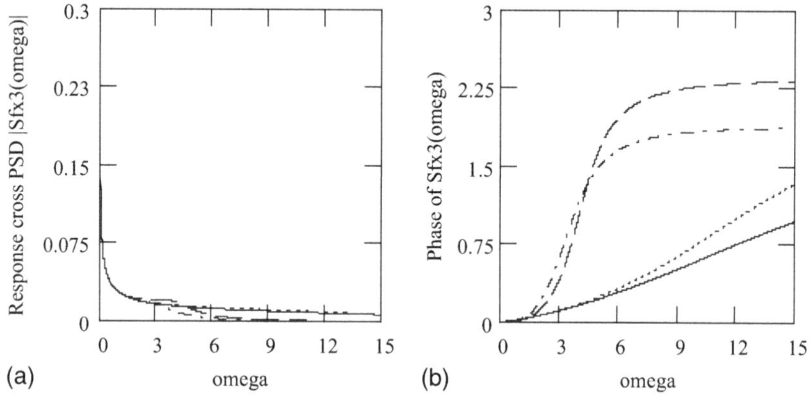


FIGURE 2.14 Cross-PSD response $S_{fx3}(\omega)$ when $m=1$, $c=0.1$, $k=36$, and $H=0.75$ for $(\alpha, \beta)=(1.3, 0.8)$ (solid), (1.3, 1.8) (dot), (2.5, 0.8) (dash), (2.8, 1.8) (dash dot). (a). $|S_{fx3}(\omega)|$. (b). Phase of $S_{fx3}(\omega)$.

## 2.5 RESPONSES OF CLASS IV FRACTIONAL VIBRATORS DRIVEN BY FGN

### 2.5.1 Computations

The motion equation of a class IV fractional vibrator is given by

$$m\frac{d^{\alpha}x_4(t)}{dt^{\alpha}}+k\frac{d^{\lambda}x_4(t)}{dt^{\lambda}}=f(t). \tag{2.23}$$

In (2.23), $x_4(t)$ is the response of a class IV fractional vibrator.

**Theorem 2.7 (PSD response IV)**

Let $S_{xx4}(\omega)$ be the PSD of $x_4(t)$. Then,

$$S_{xx4}(\omega)=\frac{V_H\sin(H\pi)\Gamma(2H+1)|\omega|^{1-2H}}{\left[k^2\omega^{2\lambda}\cos^2\frac{\lambda\pi}{2}\begin{bmatrix}\left(1-\gamma^2\dfrac{-\omega^{\alpha-2}\cos\frac{\alpha\pi}{2}}{\omega^{\lambda}\cos\frac{\lambda\pi}{2}}\right)^2\\+4\left(\gamma\dfrac{m\omega^{\alpha-1}\sin\frac{\alpha\pi}{2}+k\omega^{\lambda-1}\sin\frac{\lambda\pi}{2}}{2\sqrt{mk\omega^{\alpha+\lambda-2}\left|\cos\frac{\alpha\pi}{2}\right|\left|\cos\frac{\lambda\pi}{2}\right|}}\sqrt{\dfrac{-\omega^{\alpha-2}\cos\frac{\alpha\pi}{2}}{\omega^{\lambda}\cos\frac{\lambda\pi}{2}}}\right)^2\end{bmatrix}\right]}. \tag{2.24}$$

*Proof.* Note that $S_{xx4}(\omega) = S_{ff}(\omega)|H_4(\omega)|^2$. Following Chapter 7 in Volume I or Li [1] or [2]), $H_4(\omega)$ is given by

$$H_4(\omega)=\frac{1}{k\omega^{\lambda}\cos\frac{\lambda\pi}{2}\left(\begin{matrix}1-\gamma^2\dfrac{-\omega^{\alpha-2}\cos\frac{\alpha\pi}{2}}{\omega^{\lambda}\cos\frac{\lambda\pi}{2}}\\+i2\gamma\dfrac{m\omega^{\alpha-1}\sin\frac{\alpha\pi}{2}+k\omega^{\lambda-1}\sin\frac{\lambda\pi}{2}}{2\sqrt{mk\omega^{\alpha+\lambda-2}\left|\cos\frac{\alpha\pi}{2}\right|\left|\cos\frac{\lambda\pi}{2}\right|}}\sqrt{\dfrac{-\omega^{\alpha-2}\cos\frac{\alpha\pi}{2}}{\omega^{\lambda}\cos\frac{\lambda\pi}{2}}}\end{matrix}\right)}. \tag{2.25}$$

Thus, (2.24) holds. The proof ends.

**Theorem 2.8 (cross-PSD response IV)**

Denote by $S_{fx4}(\omega)$ the cross PSD between $f(t)$ and $x_4(t)$. Then,

$$S_{fx4}(\omega)=\frac{V_H\sin(H\pi)\Gamma(2H+1)|\omega|^{1-2H}}{k\omega^{\lambda}\cos\frac{\lambda\pi}{2}\left(\begin{matrix}1-\gamma^2\dfrac{-\omega^{\alpha-2}\cos\frac{\alpha\pi}{2}}{\omega^{\lambda}\cos\frac{\lambda\pi}{2}}\\+i2\gamma\dfrac{m\omega^{\alpha-1}\sin\frac{\alpha\pi}{2}+k\omega^{\lambda-1}\sin\frac{\lambda\pi}{2}}{2\sqrt{mk\omega^{\alpha+\lambda-2}\left|\cos\frac{\alpha\pi}{2}\right|\left|\cos\frac{\lambda\pi}{2}\right|}}\sqrt{\dfrac{-\omega^{\alpha-2}\cos\frac{\alpha\pi}{2}}{\omega^{\lambda}\cos\frac{\lambda\pi}{2}}}\end{matrix}\right)}. \tag{2.26}$$

*Proof.* Doing the operation of $S_{fx4}(\omega) = S_{ff}(\omega)H_4(\omega)$ and considering (2.25) yields (2.26). This finishes the proof.

Let $r_{xx4}(\tau)$ be the ACF of $x_4(t)$. Let $h_4(\tau)$ be the impulse response of a class IV fractional vibrator. Then,

$$r_{xx4}(\tau) = r_{ff}(\tau) * h_4(\tau) * h_4(-\tau). \tag{2.27}$$

Let $r_{fx4}(\tau)$ be the cross-correlation between the excitation $f(t)$ and the response $x_4(t)$. Then,

$$r_{fx4}(\tau) = r_{ff}(\tau) * h_4(\tau). \tag{2.28}$$

In (2.27) and (2.28), according to Chapter 7 in Volume I or Li [1] or [2], $h_4(t)$ is given by

$$h_4(t) = \frac{e^{-\frac{m\omega^{\alpha-1}\sin\frac{\alpha\pi}{2}+k\omega^{\lambda-1}\sin\frac{\lambda\pi}{2}}{2\sqrt{mk\omega^{\alpha+\lambda-2}\left|\cos\frac{\alpha\pi}{2}\right|\cos\frac{\lambda\pi}{2}}}\sqrt{\frac{\omega^{\lambda}\cos\frac{\lambda\pi}{2}}{-\omega^{\alpha-2}\cos\frac{\alpha\pi}{2}}}\,\omega_n t}}{\omega_n m\sqrt{-\omega^{\alpha-2}\cos\frac{\alpha\pi}{2}}\sqrt{\omega^{\lambda}\cos\frac{\lambda\pi}{2}}\sqrt{1-\left(\frac{m\omega^{\alpha-1}\sin\frac{\alpha\pi}{2}+k\omega^{\lambda-1}\sin\frac{\lambda\pi}{2}}{2\sqrt{mk\omega^{\alpha+\lambda-2}\left|\cos\frac{\alpha\pi}{2}\right|\cos\frac{\lambda\pi}{2}}}\right)^2}}$$
$$\sin\omega_n\sqrt{\frac{\omega^{\lambda}\cos\frac{\lambda\pi}{2}}{-\omega^{\alpha-2}\cos\frac{\alpha\pi}{2}}}\sqrt{1-\left(\frac{m\omega^{\alpha-1}\sin\frac{\alpha\pi}{2}+k\omega^{\lambda-1}\sin\frac{\lambda\pi}{2}}{2\sqrt{mk\omega^{\alpha+\lambda-2}\left|\cos\frac{\alpha\pi}{2}\right|\cos\frac{\lambda\pi}{2}}}\right)^2}\,tu(t). \tag{2.29}$$

## 2.5.2 Effect of $(\alpha, \lambda)$ on Responses

We illustrate some plots of $|H_4(\omega)|^2$ in Figure 2.15. Figure 2.16 indicates some plots of $S_{xx4}(\omega)$. Some plots of $S_{fx4}(\omega)$ are shown in Figure 2.17. Figures 2.16 and 2.17 exhibit that there is noticeable effect of $(\alpha, \lambda)$ on the responses of a class IV fractional vibrator excited by fGn.

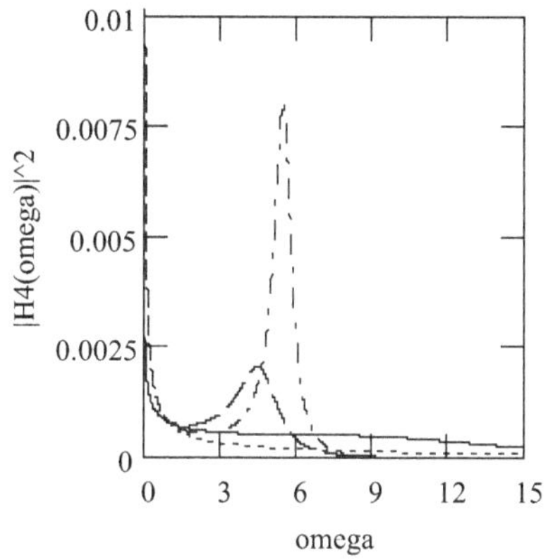


FIGURE 2.15 Plots of $|H_4(\omega)|^2$ in log when $m = 1$, $c = 0$, and $k = 36$, for $(\alpha, \lambda) =$ (1.6, 0.2) (solid), (1.6, 0.4) (dot), (2.5, 0.2) (dash), (2.5, 0.4) (dash dot).

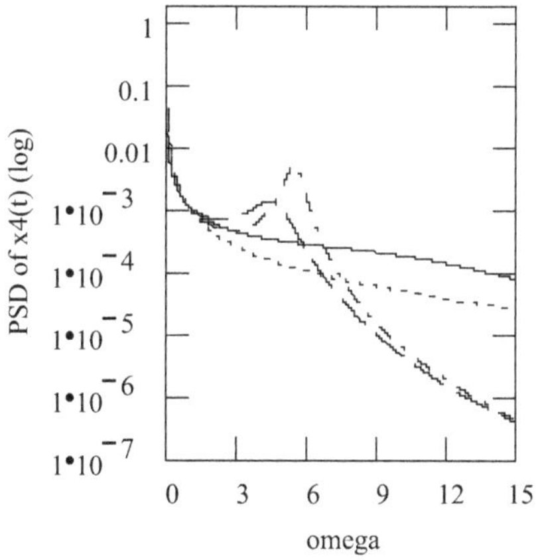


FIGURE 2.16 Response PSD $S_{xx4}(\omega)$ in log when $m = 1, c = 0, k = 36$, and $H = 0.75$ for $(\alpha, \lambda) =$ (1.6, 0.8) (solid), (1.6, 1.8) (dot), (2.5, 0.8) (dash), (2.8, 1.8) (dash dot).

## 2.6 RESPONSES OF CLASS V FRACTIONAL VIBRATORS DRIVEN BY FGN

### 2.6.1 Computation Methods

The motion equation of a class V fractional vibrator is in the form

$$m\frac{d^2 x_5(t)}{dt^2} + k\frac{d^\lambda x_5(t)}{dt^\lambda} = f(t). \tag{2.30}$$

In (2.30), $0 \le \lambda < 1$, $x_5(t)$ is the response of a class V fractional vibrator.

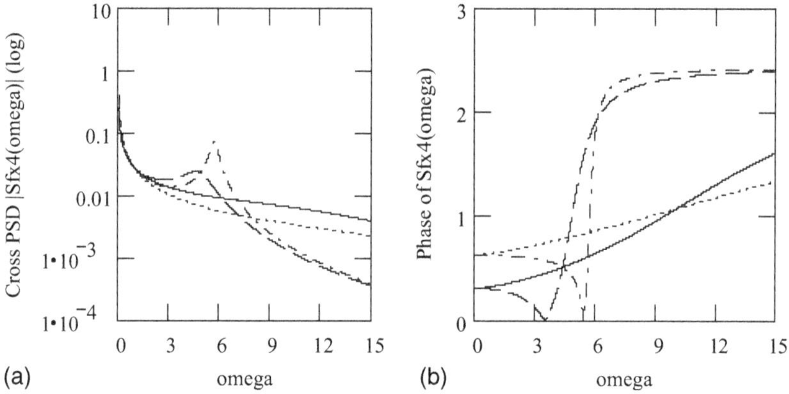


FIGURE 2.17 Cross-PSD $S_{\mathrm{fx4}}(\omega)$ when $m = 1$, $c = 0$, $k = 36$, and $V = 15$ for $(\alpha, \lambda) = (1.6, 0.2)$ (solid), $(1.6, 0.4)$ (dot), $(2.45, 0.2)$ (dash), $(2.45, 0.4)$ (dash dot). (a). $|S_{\mathrm{fx4}}(\omega)|$. (b). Phase of $S_{\mathrm{fx4}}(\omega)$.

**Theorem 2.9 (PSD response V)**

Denote by $S_{xx5}(\omega)$ the PSD of $x_5(t)$. Then,

$$S_{xx5}(\omega)=\frac{V_H\sin(H\pi)\Gamma(2H+1)|\omega|^{1-2H}}{k^2\omega^{2\lambda}\cos^2\frac{\lambda\pi}{2}\left[\left(1-\frac{\gamma^2}{\omega^{\lambda}\cos\frac{\lambda\pi}{2}}\right)^2+4\gamma^2\left(\frac{k\omega^{\lambda-1}\sin\frac{\lambda\pi}{2}}{2\sqrt{mk\omega^{\lambda}\cos\frac{\lambda\pi}{2}}}\sqrt{\frac{1}{\omega^{\lambda}\cos\frac{\lambda\pi}{2}}}\right)^2\right]}. \tag{2.31}$$

*Proof.* Because of $S_{xx5}(\omega) = S_{ff}(\omega)|H_5(\omega)|^2$, where, according to Chapter 7 in Volume I or Li [1] or [2], $H_5(\omega)$ is given by

$$H_5(\omega)=\frac{1}{k\omega^{\lambda}\cos\frac{\lambda\pi}{2}\left(1-\frac{\gamma^2}{\omega^{\lambda}\cos\frac{\lambda\pi}{2}}+i2\gamma\frac{k\omega^{\lambda-1}\sin\frac{\lambda\pi}{2}}{2\sqrt{mk\omega^{\lambda}\cos\frac{\lambda\pi}{2}}}\sqrt{\frac{1}{\omega^{\lambda}\cos\frac{\lambda\pi}{2}}}\right)}, \tag{2.32}$$

(2.31) holds. The proof ends.

**Theorem 2.10 (cross-PSD response V)**

Let $S_{fx5}(\omega)$ be the cross PSD between $f(t)$ and $x_5(t)$. Then,

$$S_{fx5}(\omega)=\frac{V_H \sin(H\pi)\Gamma(2H+1)|\omega|^{1-2H}}{k\omega^{\lambda}\cos\frac{\lambda\pi}{2}\left(1-\frac{\gamma^2}{\omega^{\lambda}\cos\frac{\lambda\pi}{2}}+i2\gamma\frac{k\omega^{\lambda-1}\sin\frac{\lambda\pi}{2}}{2\sqrt{mk\omega^{\lambda}\cos\frac{\lambda\pi}{2}}}\sqrt{\frac{1}{\omega^{\lambda}\cos\frac{\lambda\pi}{2}}}\right)}. \tag{2.33}$$

*Proof.* Due to $S_{fx5}(\omega) = S_{ff}(\omega)H_5(\omega)$ and (2.32), (2.33) holds. This finishes the proof.

Let $r_{xx5}(\tau)$ be the ACF of $x_5(t)$. Denote by $h_5(\tau)$ the impulse response of a class V fractional vibrator. Then,

$$r_{xx5}(\tau) = r_{ff}(\tau)*h_5(\tau)*h_5(-\tau). \tag{2.34}$$

Let $r_{fx5}(\tau)$ be the cross-correlation between $f(t)$ and $x_5(t)$. Then,

$$r_{fx5}(\tau) = r_{ff}(\tau)*h_5(\tau). \tag{2.35}$$

In (2.34) and (2.35), following Chapter 7 in Volume I or Li [1] or [2],

$$h_5(t)=e^{-\frac{k\omega^{\lambda-1}\sin\frac{\lambda\pi}{2}}{2\sqrt{mk\omega^{\lambda}\cos\frac{\lambda\pi}{2}}}\sqrt{\omega^{\lambda}\cos\frac{\lambda\pi}{2}}\,\omega_n t}\frac{1}{m\omega_{\mathrm{eqd5}}}\sin\omega_{\mathrm{eqd5}}tu(t). \tag{2.36}$$

Refer to Chapter 7 in Volume I or Li [1] or [2] for the expression of $\omega_{\mathrm{eqd5}}$ in (2.36).

## 2.6.2 Effect of $\lambda$ on Responses

Figure 2.18 illustrates some plots of $|H_5(\omega)|^2$. Figure 2.19 shows some plots of $S_{xx5}(\omega)$. Figure 2.20 indicates some plots of $S_{fx5}(\omega)$. Figures 2.19 and 2.20 show that the effect of the fractional order $\lambda$ on the responses of a class V fractional vibration system under the excitation of fGn is considerable.

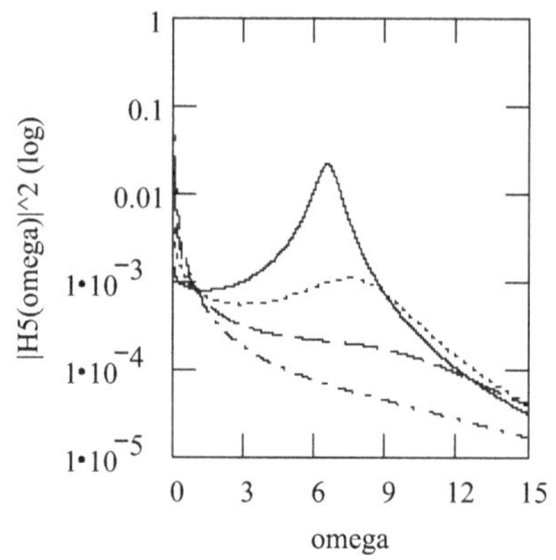


FIGURE 2.18 Plots of $|H_5(\omega)|^2$ (log) when $m = 1$, $c = 0$, and $k = 36$, for $\lambda = 0.1$ (solid), 0.3 (dot), 0.5 (dash), 0.7 (dash dot).

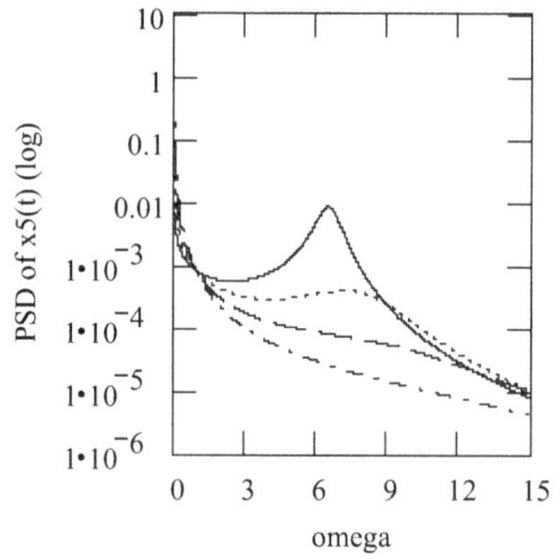


FIGURE 2.19 Response PSD $S_{xx5}(\omega)$ in log when $m = 1$, $c = 0$, $k = 36$, and $H = 0.75$ for $\lambda = 0.1$ (solid), 0.3 (dot), 0.5 (dash), 0.7 (dash dot).

## 2.7 RESPONSES OF CLASS VI FRACTIONAL VIBRATORS DRIVEN BY FGN

### 2.7.1 Computations

The motion equation of a class VI fractional vibrator is expressed by

$$m\frac{d^{\alpha}x_6(t)}{dt^{\alpha}}+c\frac{d^{\beta}x_6(t)}{dt^{\beta}}+k\frac{d^{\lambda}x_6(t)}{dt^{\lambda}}=f(t). \tag{2.37}$$

In (2.37), $x_6(t)$ is the response of a class VI fractional vibrator.

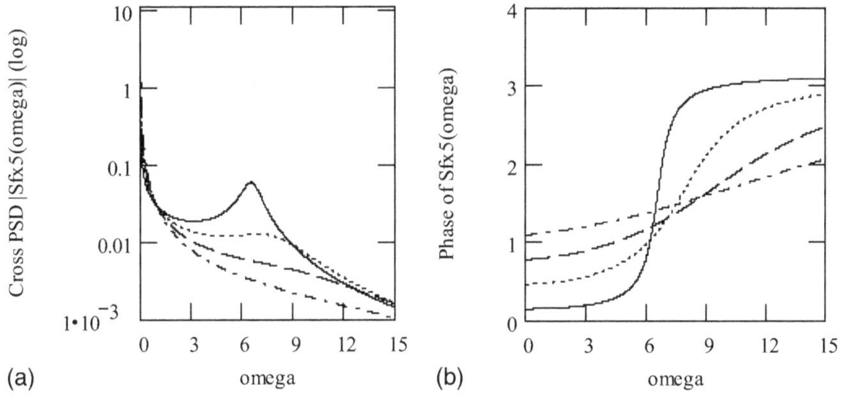


FIGURE 2.20 Cross-PSD response $S_{fx5}(\omega)$ when $m = 1$, $c = 0$, $k = 36$, and $H = 0.75$ for $\lambda = 0.1$ (solid), 0.3 (dot), 0.5 (dash), 0.7 (dash dot). (a). $|S_{fx5}(\omega)|$ in log. (b). Phase of $S_{fx5}(\omega)$.

**Theorem 2.11 (PSD response VI)**

Let $S_{xx6}(\omega)$ be the PSD of $x_6(t)$. Then,

$$S_{xx6}(\omega) = \frac{1}{k^2} \frac{V_H \sin(H\pi)\Gamma(2H+1)|\omega|^{1-2H}}{\left[\omega^{\lambda}\cos\frac{\lambda\pi}{2} + \gamma^2\left(\omega^{\alpha-2}\cos\frac{\alpha\pi}{2} + 2\varsigma\omega_n\omega^{\beta-2}\cos\frac{\beta\pi}{2}\right)\right]^2 + \gamma^2\left(\omega^{\alpha-1}\sin\frac{\alpha\pi}{2} + 2\zeta\omega_n\omega^{\beta-1}\sin\frac{\beta\pi}{2} + \omega_n^2\omega^{\lambda-1}\sin\frac{\lambda\pi}{2}\right)^2}. \tag{2.38}$$

*Proof.* According to Chapter 7 in Volume I or Li [1] or [2], $H_6(\omega)$ is given by

$$H_6(\omega) = \frac{1}{k\left[\omega^{\lambda}\cos\frac{\lambda\pi}{2} + \gamma^2\left(\omega^{\alpha-2}\cos\frac{\alpha\pi}{2} + 2\varsigma\omega_n\omega^{\beta-2}\cos\frac{\beta\pi}{2}\right) + i\gamma\left(\omega^{\alpha-1}\sin\frac{\alpha\pi}{2} + 2\zeta\omega_n\omega^{\beta-1}\sin\frac{\beta\pi}{2} + \omega_n^2\omega^{\lambda-1}\sin\frac{\lambda\pi}{2}\right)\right]}. \tag{2.39}$$

Doing the operation of $S_{xx6}(\omega) = S_{ff}(\omega)|H_6(\omega)|^2$ yields (2.38). This finishes the proof.

**Theorem 2.12 (cross-PSD response VI)**

Denote by $S_{fx6}(\omega)$ the cross PSD between $f(t)$ and $x_6(t)$. Then,

$$S_{fx6}(\omega)=\frac{V_H \sin(H\pi)\Gamma(2H+1)|\omega|^{1-2H}}{k\left[\begin{array}{l}\omega^{\lambda}\cos\frac{\lambda\pi}{2}+\gamma^2\left(\omega^{\alpha-2}\cos\frac{\alpha\pi}{2}+2\varsigma\omega_n\omega^{\beta-2}\cos\frac{\beta\pi}{2}\right)\\+i\gamma\left(\omega^{\alpha-1}\sin\frac{\alpha\pi}{2}+2\zeta\omega_n\omega^{\beta-1}\sin\frac{\beta\pi}{2}+\omega_n^2\omega^{\lambda-1}\sin\frac{\lambda\pi}{2}\right)\end{array}\right]}. \tag{2.40}$$

*Proof.* Doing the operation of $S_{fx6}(\omega)=S_{ff}(\omega)H_6(\omega)$ and taking into account (2.39) results in (2.40). The proof ends.

Let $r_{xx6}(\tau)$ be the ACF of $x_6(t)$. Denote by $h_6(\tau)$ the impulse of a class VI fractional vibrator. Applying the Wiener-Khinchin relation and the Wiener-Lee relation and the convolution theory to $S_{xx6}(\omega)=S_{ff}(\omega)|H_6(\omega)|^2$ and $S_{fx6}(\omega)=S_{ff}(\omega)H_6(\omega)$ produces the ACF response in the form

$$r_{xx6}(\tau)=r_{ff}(\tau)*h_6(\tau)*h_6(-\tau), \tag{2.41}$$

and the cross-correlation response $r_{fx6}(\tau)$ given by

$$r_{fx6}(\tau)=r_{ff}(\tau)*h_6(\tau). \tag{2.42}$$

In (2.41) and (2.42) (Chapter 7 in Volume I or Li [1] or [2]),

$$h_6(t)=e^{-\varsigma_{\mathrm{eq6}}\omega_{\mathrm{eqn6}}t}\frac{1}{m_{\mathrm{eq6}}\omega_{\mathrm{eqd6}}}\sin\omega_{\mathrm{eqd6}}tu(t). \tag{2.43}$$

Refer to Chapter 7 in Volume I or Li [1] or [2] for the expressions of $\zeta_{\mathrm{eq6}}$, $m_{\mathrm{eq6}}$, $\omega_{\mathrm{eqn6}}$, and $\omega_{\mathrm{eqd6}}$ in (2.43).

### 2.7.2 Effect of $(\alpha, \beta, \lambda)$ on Responses

Figure 2.21 demonstrates some plots of $|H_6(\omega)|^2$. Figure 2.22 indicates some plots of $S_{xx6}(\omega)$. Figure 2.23 illustrates some plots of $S_{fx6}(\omega)$. Figures 2.22 and 2.23 show that the effect of $(\alpha, \beta, \lambda)$ on the responses of a class VI fractional vibrator driven by fGn is significant.

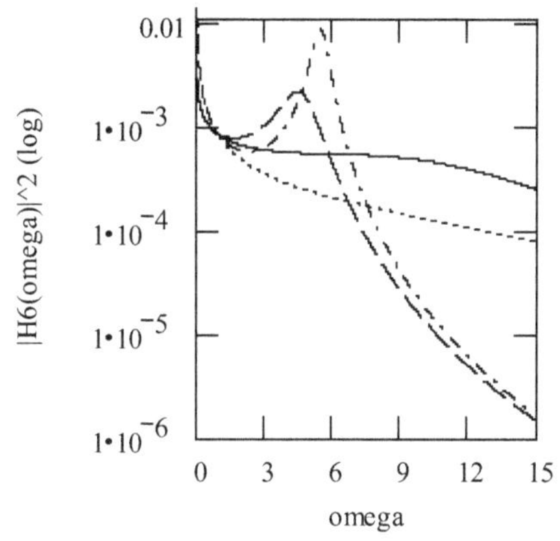


FIGURE 2.21 Plots of $|H_6(\omega)|^2$ (log) when $m = 1$, $c = 0.1$, and $k = 36$, for $(\alpha, \beta, \lambda) = (1.6, 0.5, 0.2)$ (solid), $(1.6, 1.5, 0.4)$ (dot), $(2.5, 0.5, 0.2)$ (dash), $(2.5, 0.5, 0.4)$ (dash dot).

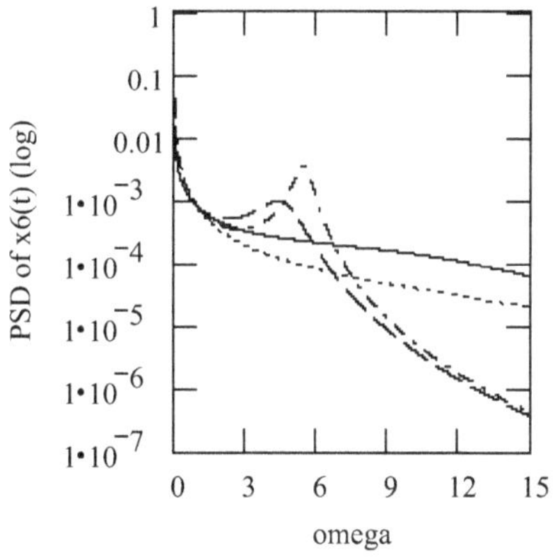


FIGURE 2.22 Response PSD $S_{xx6}(\omega)$ in log when $m = 1$, $c = 0.1$, $k = 36$, and $H = 0.75$ for $(\alpha, \beta, \lambda) = (1.6, 0.8, 0.2)$ (solid), $(1.6, 1.8, 0.4)$ (dot), $(2.5, 0.4, 0.2)$ (dash), $(2.5, 0.8, 0.4)$ (dash dot).

## 2.8 RESPONSES OF CLASS VII FRACTIONAL VIBRATORS DRIVEN BY FGN

### 2.8.1 Computations

The motion equation of a class VII fractional vibrator is given by

$$m\frac{d^2x_7(t)}{dt^2}+c\frac{d^\beta x_7(t)}{dt^\beta}+k\frac{d^\lambda x_7(t)}{dt^\lambda}=f(t),\quad 0<\beta<2,\ 0\le\lambda<1. \tag{2.44}$$

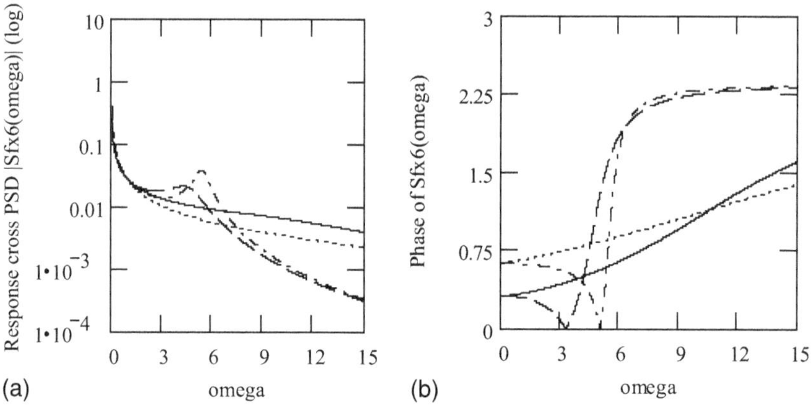


FIGURE 2.23 Cross-PSD response $S_{fx6}(\omega)$ when $m = 1, c = 0.1, k = 36$, and $H = 0.75$ for $(\alpha, \beta, \lambda) = (1.6, 0.5, 0.2)$ (solid), (1.6, 1.5, 0.4) (dot), (2.5, 0.5, 0.2) (dash), (2.5, 0.5, 0.4) (dash dot). (a). $|S_{fx6}(\omega)|$ in log. (b). Phase of $S_{fx6}(\omega)$.

In (2.44), $x_7(t)$ is the response of a class VII fractional vibrator.

**Theorem 2.13 (PSD response VII)**

Denote by $S_{xx7}(\omega)$ the PSD of $x_7(t)$. Then,

$$S_{xx7}(\omega) = \frac{1}{k^2} \frac{V_H \sin(H\pi)\Gamma(2H+1)|\omega|^{1-2H}}{\left[\omega^\lambda \cos\frac{\lambda\pi}{2} - \gamma\left(1 - 2\varsigma\omega_n\omega^{\beta-2}\cos\frac{\beta\pi}{2}\right)\right]^2 + \gamma^2\left(2\varsigma\omega^{\beta-1}\sin\frac{\beta\pi}{2} + \omega_n\omega^{\lambda-1}\sin\frac{\lambda\pi}{2}\right)^2}. \tag{2.45}$$

*Proof.* Note that $S_{xx7}(\omega) = S_{ff}(\omega)|H_7(\omega)|^2$. Besides (Chapter 7 in Volume I or Li [2]), $H_7(\omega)$ is given by (2.46) in the form

$$H_7(\omega) = \frac{1}{k\left[\begin{array}{l}\omega^\lambda \cos\frac{\lambda\pi}{2} - \gamma\left(1 - 2\varsigma\omega_n\omega^{\beta-2}\cos\frac{\beta\pi}{2}\right)\\ +i\gamma\left(2\varsigma\omega^{\beta-1}\sin\frac{\beta\pi}{2} + \omega_n\omega^{\lambda-1}\sin\frac{\lambda\pi}{2}\right)\end{array}\right]}. \tag{2.46}$$

Therefore, (2.45) holds. The proof ends.

**Theorem 2.14 (cross-PSD response VII)**

Let $S_{fx7}(\omega)$ be the cross PSD between $f(t)$ and $x_7(t)$. Then,

$$S_{fx7}(\omega)=\frac{V_H\sin(H\pi)\Gamma(2H+1)|\omega|^{1-2H}}{k\left[\begin{array}{l}\omega^{\lambda}\cos\frac{\lambda\pi}{2}-\gamma\left(1-2\varsigma\omega_n\omega^{\beta-2}\cos\frac{\beta\pi}{2}\right)\\+i\gamma\left(2\varsigma\omega^{\beta-1}\sin\frac{\beta\pi}{2}+\omega_n\omega^{\lambda-1}\sin\frac{\lambda\pi}{2}\right)\end{array}\right]}. \tag{2.47}$$

*Proof.* Performing $S_{fx7}(\omega) = S_{ff}(\omega)H_7(\omega)$ and considering (2.46) produces (2.47). The proof is finished.

Let $r_{xx7}(\tau)$ be the ACF of $x_7(t)$. Denote by $h_7(\tau)$ the impulse response of a class VII fractional vibrator. Then, the ACF response $r_{xx7}(\tau)$ is in the form

$$r_{xx7}(\tau) = r_{ff}(\tau)*h_7(\tau)*h_7(-\tau). \tag{2.48}$$

Let $r_{fx7}(\tau)$ be the cross-correlation between $f(t)$ and $x_7(t)$. Then,

$$r_{fx7}(\tau) = r_{ff}(\tau)*h_7(\tau). \tag{2.49}$$

In (2.48) and (2.49) (Chapter 7 in Volume I or Li [2]),

$$h_7(t)=e^{-\varsigma_{eq7}\omega_{eqn7}t}\frac{1}{m_{eq7}\omega_{eqd7}}\sin\omega_{eqd7}t,\quad t\geq 0. \tag{2.50}$$

Refer to Chapter 7 in Volume I or Li [2] for the expressions of $\zeta_{eq7}$, $m_{eq7}$, $\omega_{eqn7}$, and $\omega_{eqd7}$ in (2.50).

### 2.8.2 Effect of ($\beta$, $\lambda$) on Responses

Figure 2.24 indicates some plots of $|H_7(\omega)|^2$, Figure 2.25 shows some plots of $S_{xx7}(\omega)$, and Figure 2.26 illustrates some plots of $S_{fx7}(\omega)$. Figures 2.25 and 2.26 indicate that the effect of ($\beta$, $\lambda$) on the responses of class VII fractional vibrators excited by fGn is significant.

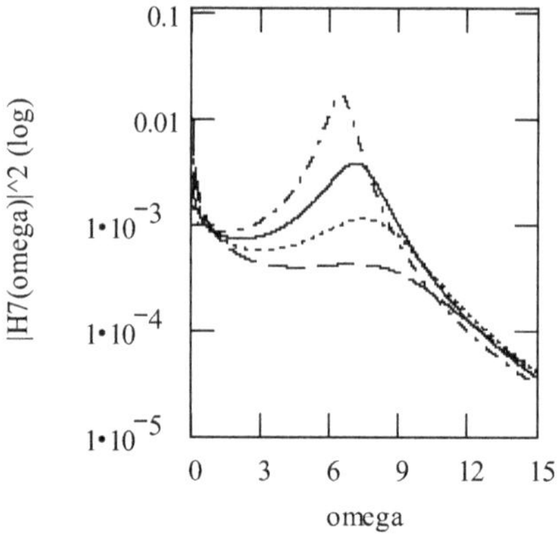


FIGURE 2.24 Plots of $|H_7(\omega)|^2$ (log) when $m = 1$, $c = 0.1$, and $k = 36$, for $(\beta, \lambda) =$ (0.5, 0.2) (solid), (0.5, 0.3) (dot), (1.5, 0.4) (dash), (1.5, 0.1) (dash dot).

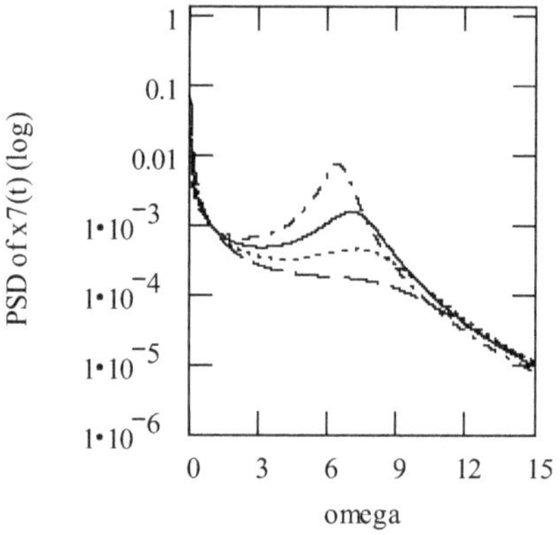


FIGURE 2.25 Response PSD $S_{xx7}(\omega)$ in log when $m = 1$, $c = 0.1$, $k = 36$, and $H =$ 0.75 for $(\beta, \lambda) =$ (0.5, 0.2) (solid), (0.5, 0.3) (dot), (1.5, 0.4) (dash), (1.5, 0.1) (dash dot).

## 2.9 SUMMARY

We have presented the analytic expressions of the PSD and cross-PSD responses of seven classes of fractional vibrators driven by fGn. We have shown that there are noticeable effects of fractional orders, say, $\alpha$ or $\beta$ or $\lambda$ on responses. In addition, the statistical dependences (long-range dependence or short-range dependence) of the responses rely on the excitation fGn because $S_{xxi}(0) = \infty$ when $0.5 < H < 1$ or $S_{xxi}(0) < \infty$ when $0 < H < 0.5$ for $i = 1, \ldots, 7$.

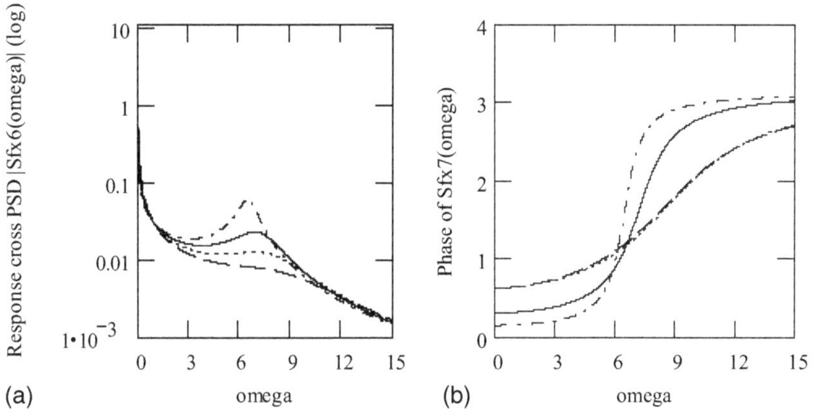


FIGURE 2.26 Cross-PSD response $S_{fx7}(\omega)$ when $m = 1$, $c = 0.1$, $k = 36$, and $H = 0.75$ for $(\beta, \lambda) = (0.5, 0.2)$ (solid), $(0.5, 0.3)$ (dot), $(1.5, 0.4)$ (dash), $(1.5, 0.1)$ (dash dot). (a). $|S_{fx7}(\omega)|$. (b). Phase of $S_{fx7}(\omega)$.

## 2.10 EXERCISES

2.1. Denote the PSD of fGn as

$$S_{ff}(\omega) = V_H \sin(H\pi)\Gamma(2H+1)|\omega|^{1-2H},$$

where $0 < H < 1$ and $V_H = \Gamma(1-2H)\dfrac{\cos \pi H}{\pi H}$. Find the inverse Fourier transform of $S_{ff}(\omega)$.

2.2. Let

$$h_1(t) = \frac{e^{-\frac{\omega \sin\frac{\alpha\pi}{2}}{2\left|\cos\frac{\alpha\pi}{2}\right|}t} \sin\left(\dfrac{\omega_n}{\sqrt{\omega^{\alpha-2}\left|\cos\frac{\alpha\pi}{2}\right|}}\sqrt{1-\dfrac{\omega^{2\alpha}\sin^2\frac{\alpha\pi}{2}}{4\omega_n^2\left|\cos\frac{\alpha\pi}{2}\right|}}\,t\right)}{m\omega_n\sqrt{\omega^{\alpha-2}\left|\cos\frac{\alpha\pi}{2}\right|}\sqrt{1-\dfrac{\omega^{2\alpha}\sin^2\frac{\alpha\pi}{2}}{4\omega_n^2\left|\cos\frac{\alpha\pi}{2}\right|}}}u(t),$$

where $u(t)$ is the unit step function and $0 < \alpha < 3$. Find

$$r_{xx1}(\tau) = r_{ff}(\tau) * h_1(\tau) * h_1(-\tau),$$

and

$$r_{fx1}(\tau) = r_{ff}(\tau) * h_1(\tau),$$

where $r_{ff}(\tau) = \mathrm{F}^{-1}[S_{ff}(\omega)]$.

2.3. Find $\int_{-\infty}^{\infty} r_{xx1}(\tau)d\tau$ for $0 < H < 0.5$.

2.4. Find $\int_{-\infty}^{\infty} r_{xx1}(\tau)d\tau$ for $0.5 < H < 1$.

2.5. Let

$$h_3(t) = \frac{e^{-\frac{m\omega^{\alpha-1}\sin\frac{\alpha\pi}{2} + c\omega^{\beta-1}\sin\frac{\beta\pi}{2}}{2\sqrt{-\left(m\omega^{\alpha-2}\cos\frac{\alpha\pi}{2} + c\omega^{\beta-2}\cos\frac{\beta\pi}{2}\right)k}}\omega_{\mathrm{eqn3}}t} \sin\omega_{\mathrm{eqd3}}t}{-\left(m\omega^{\alpha-2}\cos\frac{\alpha\pi}{2} + c\omega^{\beta-2}\cos\frac{\beta\pi}{2}\right)\omega_{\mathrm{eqd3}}} u(t).$$

Find $r_{xx3}(\tau) = r_{ff}(\tau) * h_3(\tau) * h_3(-\tau)$, where $r_{ff}(\tau) = \mathrm{F}^{-1}[S_{ff}(\omega)]$.

2.6. Find $r_{fx3}(\tau) = r_{ff}(\tau) * h_3(\tau)$, where $r_{ff}(\tau) = \mathrm{F}^{-1}[S_{ff}(\omega)]$.

2.7. Prove $\int_{-\infty}^{\infty} r_{xx3}(\tau)d\tau < \infty$ for $0 < H < 0.5$.

2.8. Prove $\int_{-\infty}^{\infty} r_{xx3}(\tau)d\tau = \infty$ for $0.5 < H < 1$.

## REFERENCES


1. M. Li, *Fractional Vibrations with Applications to Euler-Bernoulli Beams*, CRC Press, Boca Raton, 2023.
2. M. Li, Analytic theory of seven classes of fractional vibrations based on elementary functions: A tutorial review, *Symmetry*, 16(9): 2024, 1202.
3. X. Sun, M. Li, and W. Zhao, Moderate deviations for stochastic fractional heat equation driven by fractional noise, *Complexity*, 2018: 2018, 7402764, 17.
4. X. Sun, R. Guo, and M. Li, Some properties of bifractional Bessel processes driven by bifractional Brownian motion, *Mathematical Problems in Engineering*, 2020: 2020, 7037602, 13.
5. W. Wang, Z. Yan, and X. Liu, The escape problem and stochastic resonance in a bistable system driven by fractional Gaussian noise, *Physics Letters A*, 381(29): 2017, 2324–2336.
6. Y. Hu and X. Y. Zhou, Stochastic control for linear systems driven by fractional noises, *SIAM Journal on Control & Optimization*, 43(6): 2005, 2245–2277.

7. Y. Liu, Y. Wang, and T. Caraballo, The continuity, regularity and polynomial stability of mild solutions for stochastic 2D-Stokes equations with unbounded delay driven by tempered fractional Gaussian noise, *Stochastics and Dynamics*, 22(5): 2022, 2250022.
8. M. Li, *Fractal Teletraffic Modeling and Delay Bounds in Computer Communications*, CRC Press, Boca Raton, 2022.
9. M. Li, *Multi-Fractal Traffic and Anomaly Detection in Computer Communications*, CRC Press, Boca Raton, 2022.
10. M. Li, Fractal time series—a tutorial review, *Mathematical Problems in Engineering*, 2010: 2010, 157264, 26.
11. M. Li and S. C. Lim, A rigorous derivation of power spectrum of fractional Gaussian noise, *Fluctuation and Noise Letters*, 6(4): 2006, C33–C36.
12. M. Li, Generation of teletraffic of generalized Cauchy type, *Physica Scripta*, 81(2): 2010, 025007, 10.
13. M. Li and C.-H. Chi, A correlation-based computational method for simulating long-range dependent data, *Journal of the Franklin Institute*, 340(6–7): 2003, 503–514.
14. M. Li, PSD and cross PSD of responses of seven classes of fractional vibrations driven by fGn, fBm, fractional OU process, and von Kármán process, *Symmetry*, 16(5): 2024, 635.

CHAPTER 3

# Responses of Fractional Vibrations Driven by Generalized Fractional Gaussian Noise

THE CONTRIBUTIONS IN THIS chapter are in three aspects. One is to propose the closed-form analytic expressions of the power spectrum density (PSD) responses and cross-PSD responses to seven classes of fractional vibrators under the excitation of generalized fractional Gaussian noise (gfGn). The other is to show that orders of fractional vibration systems have noticeable effects on the responses. Besides, we show that the statistical dependences (long-range or short range) of responses follow those of gfGn.

## 3.1 BACKGROUND

Recently, Li introduced the analytic theory of seven classes of fractional vibrators using elementary functions (Chapters 6–7 in Volume I, and Li [1–3]). On the other side, Li brought forward a kind of fractional process called the generalized fractional Gaussian noise (gfGn) (Li [4–7]). That was recently paid attention to in applications, see, for example, Pinchas and Avraham [8], Sousa-Vieira and Fernández-Veiga [9], Gorev et al. [10], Sheluhin et al. [11], Starchenko [12]. However, what are analytic expressions of


DOI: 10.1201/9781003657903-3

responses to seven classes of fractional vibration systems driven by gfGn is an open problem. This chapter gives a solution to that problem.

The rest of the chapter is organized as follows. In Sections 3.2–3.8, we present the analytic expressions of the PSD and cross-PSD responses to seven classes of fractional vibration systems driven by gfGn. The summary is given in Section 3.9.

## 3.2 RESPONSES OF CLASS I FRACTIONAL VIBRATORS DRIVEN BY GFGN

### 3.2.1 Computations

The motion equation of a class I fractional vibrator is in the form

$$m\frac{d^{\alpha}x_1(t)}{dt^{\alpha}}+k\frac{dx_1(t)}{dt}=f(t). \tag{3.1}$$

In (3.1), $1 < \alpha < 3$, $x_1(t)$ is the response of a class I fractional vibrator, $f(t)$ is the driven force, and $m$ and $k$ are the primary mass and stiffness, respectively.

Let $f(t)$ be gfGn in what follows. Let $S_{ff}(\omega)$ be the PSD of $f(t)$. Then (Li [4, 5]),

$$S_{ff}(\omega)\approx -V_H H(2H-1)\sin\left[a\pi(H-1)\right]\Gamma(2aH-2a+1)\left|\omega\right|^{-2a(H-1)-1}. \tag{3.2}$$

In (3.2), $H \in (0, 1)$ is the Hurst parameter, $0 < a \le 1$, $V_H=\Gamma(1-2H)\dfrac{\cos\pi H}{\pi H}$. Figure 3.1 indicates some plots of $S_{ff}(\omega)$.

**Theorem 3.1 (PSD response I)**

Denote by $S_{xx1}(\omega)$ the PSD of $x_1(t)$. Then,

$$S_{xx1}(\omega)=\frac{-V_H H(2H-1)\sin\left[a\pi(H-1)\right]\Gamma(2aH-2a+1)\left|\omega\right|^{-2a(H-1)-1}}{k^2\left[\left(1-\left|\dfrac{\omega^{\alpha}}{\omega_n^2}\right|\cos\dfrac{\alpha\pi}{2}\right)^2+\left(\dfrac{\omega^{\alpha}}{\omega_n^2}\sin\dfrac{\alpha\pi}{2}\right)^2\right]}, \tag{3.3}$$

where $\omega_n^2=\dfrac{k}{m}$.

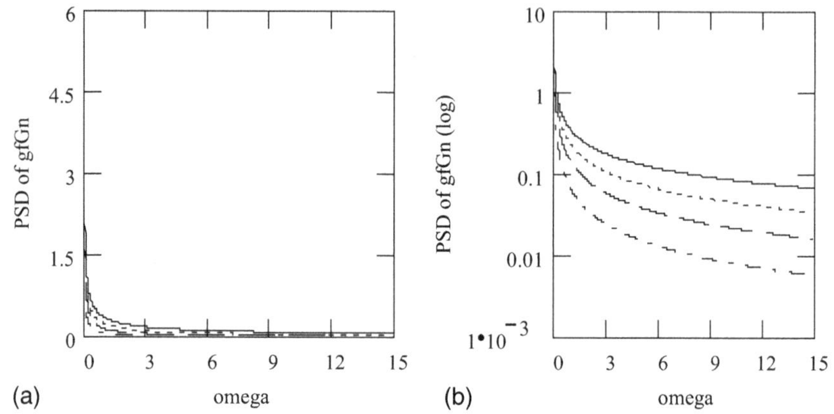


FIGURE 3.1 Plots of PSD of gfGn for $(a, H) = (0.8, 0.75)$ (solid), $(0.6, 0.75)$ (dot), $(0.4, 0.75)$ (dash), $(0.2, 0.75)$ (dash dot). (a). PSD in ordinary coordinate system. (b). PSD in log.

*Proof.* Note that $S_{xx1}(\omega) = S_{ff}(\omega)|H_1(\omega)|^2$. As $H_1(\omega)$ is given by (Chapter 7 in Volume I or Li [1–3])

$$H_1(\omega) = \frac{1}{k\left(1 - \frac{\omega^\alpha}{\omega_n^2}\left|\cos\frac{\alpha\pi}{2}\right| + i\frac{\omega^\alpha}{\omega_n^2}\sin\frac{\alpha\pi}{2}\right)}, \tag{3.4}$$

(3.3) holds. This finishes the proof.

**Theorem 3.2 (cross-PSD response I)**

Let $S_{fx1}(\omega)$ be the cross PSD between $f(t)$ and $x_1(t)$. Then,

$$S_{fx1}(\omega) = \frac{-V_H H(2H-1)\sin\left[a\pi(H-1)\right]\Gamma(2aH-2a+1)|\omega|^{-2a(H-1)-1}}{k\left(1 - \frac{\omega^\alpha}{\omega_n^2}\left|\cos\frac{\alpha\pi}{2}\right| + i\frac{\omega^\alpha}{\omega_n^2}\sin\frac{\alpha\pi}{2}\right)}. \tag{3.5}$$

*Proof.* Doing the operation of $S_{fx1}(\omega) = S_{ff}(\omega)H_1(\omega)$ with the consideration of (3.4) yields (3.5). The proof ends.

Let $r_{xx1}(\tau)$ be the autocorrelation function (ACF) of $x_1(t)$. Denote by $h_1(\tau)$ the impulse response of a class I fractional vibrator. Then, the ACF response $r_{xx1}(\tau)$ is given by

$$r_{xx1}(\tau) = r_{ff}(\tau) * h_1(\tau) * h_1(-\tau). \tag{3.6}$$

In (3.6), $*$ stands for the convolution operation. Denote by $r_{fx1}(\tau)$ the cross-correlation between $f(t)$ and $x_1(t)$. Then, the cross-correlation response $r_{fx1}(\tau)$ is expressed by

$$r_{fx1}(\tau) = r_{ff}(\tau) * h_1(\tau). \tag{3.7}$$

In (3.6) and (3.7) (Chapter 7 in Volume I, Li [1–3]),

$$h_1(t) = \frac{e^{-\frac{\omega \sin\frac{\alpha\pi}{2}}{2\left|\cos\frac{\alpha\pi}{2}\right|}t} \sin\left(\frac{\omega_n}{\sqrt{\omega^{\alpha-2}\left|\cos\frac{\alpha\pi}{2}\right|}}\sqrt{1-\frac{\omega^{2\alpha}\sin^2\frac{\alpha\pi}{2}}{4\omega_n^2\left|\cos\frac{\alpha\pi}{2}\right|}}\,t\right)}{m\omega_n\sqrt{\omega^{\alpha-2}\left|\cos\frac{\alpha\pi}{2}\right|}\sqrt{1-\frac{\omega^{2\alpha}\sin^2\frac{\alpha\pi}{2}}{4\omega_n^2\left|\cos\frac{\alpha\pi}{2}\right|}}}u(t). \tag{3.8}$$

In (3.8), $u(t)$ is the unit step function.

### 3.2.2 Effect of $\alpha$ on Responses

Figure 3.2 indicates some plots of $|H_1(\omega)|^2$. Some plots of $S_{xx1}(\omega)$ are shown in Figure 3.3. When $\alpha = 2$, a class I fractional vibrator becomes a

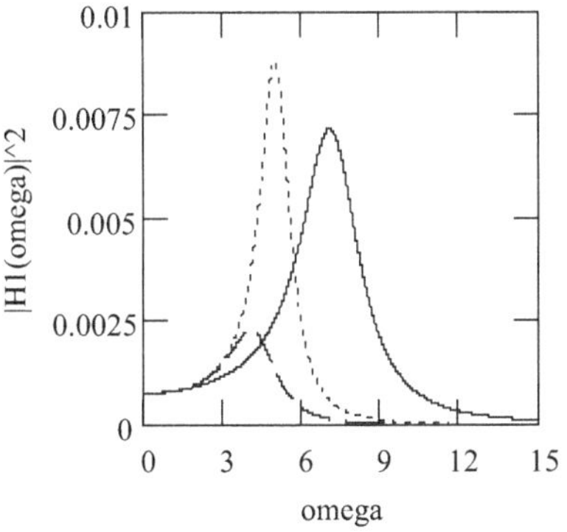


FIGURE 3.2 Plots of $|H_1(\omega)|^2$ with $\alpha = 1.8$ (solid), 2.4 (dot), 2.8 (dash) when $m = 1$ and $k = 36$.

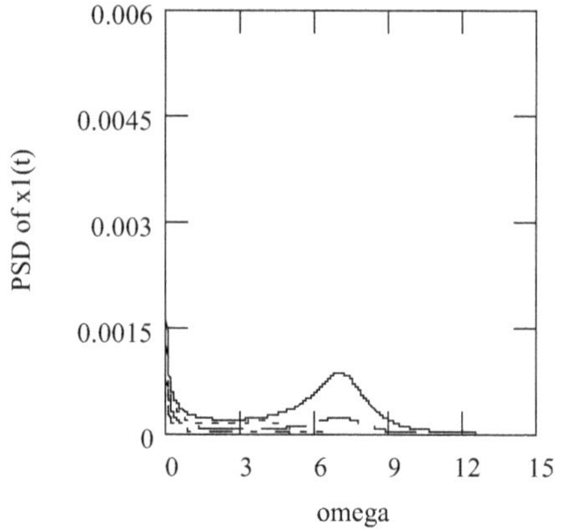


FIGURE 3.3 Plots of response PSD $S_{xx1}(\omega)$ for $m = 1$ and $k = 36$ when $(\alpha, a, H) =$ (1.8, 0.8, 0.75) (solid), (2.2, 0.6, 0.75) (dot), (2.4, 0.4, 0.75) (dash), (2.4, 0.2, 0.75) (dash dot).

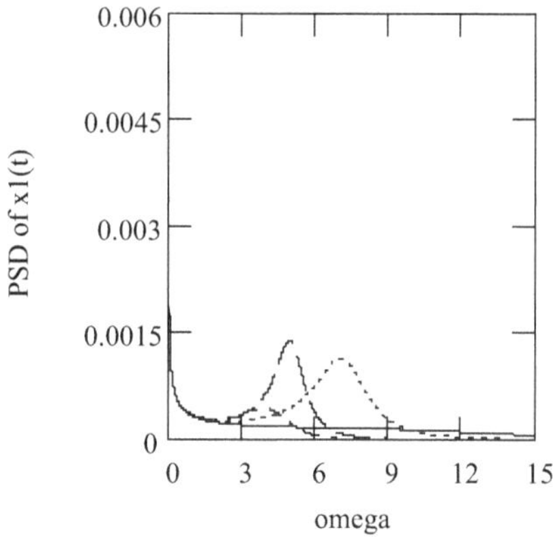


FIGURE 3.4 Observing $\alpha$ effect on $S_{xx1}(\omega)$ for $(a, H) = (0.9, 0.75)$ when $\alpha = 1.4$ (solid), 1.8 (dot), 2.2 (dash), 2.4 (dash dot).

conventional damping-free vibrator. In general, there are three parameters affecting response $S_{xx1}(\omega)$. From a view of system, the parameter is $\alpha$. From a view of signals, the parameters are $a$ and $H$. Figure 3.4 shows the effect of $\alpha$ on $S_{xx1}(\omega)$. Figure 3.5 illustrates some plots of $|S_{fx1}(\omega)|$. Some plots of driven gfGn and the response $x_1(t)$ are shown in Figure 3.6. Figures 3.5 and 3.6 exhibit that there is noticeable effect of the fractional order $\alpha$ on the responses to class I fractional vibration systems driven by gfGn.

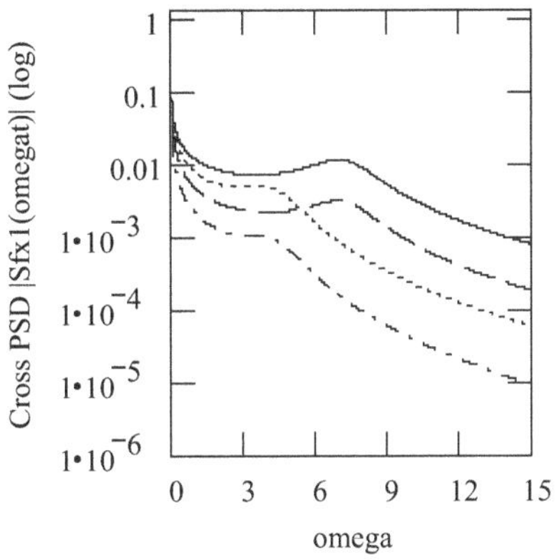


FIGURE 3.5 Illustrations of cross PSD $|S_{fx1}(\omega)|$ in log for $m = 1$ and $k = 36$ when $(\alpha, a, H) = (1.8, 0.8, 0.75)$ (solid), (2.4, 0.6, 0.75) (dot), (1.8, 0.4, 0.75) (dash), (2.4, 0.2, 0.75) (dash dot).

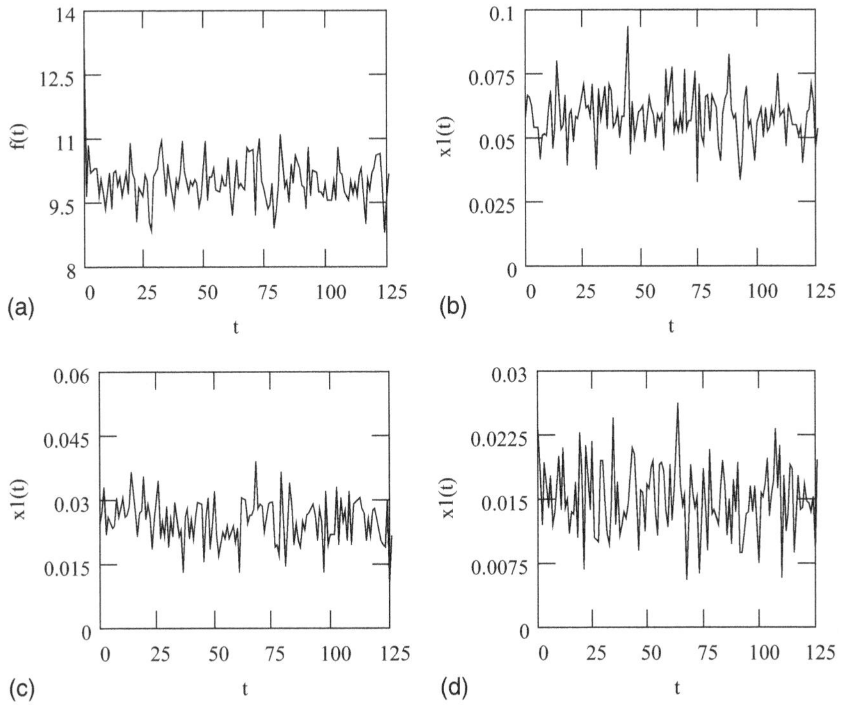


FIGURE 3.6 Plots of gfGn and $x_1(t)$ when $m = 1$ and $k = 36$. (a). Driven gfGn for $(a, H) = (0.8, 0.75)$. (b). Response $x_1(t)$ for $(\alpha, a, H) = (1.8, 0.8, 0.75)$. (c). Response $x_1(t)$ for $(\alpha, a, H) = (2.4, 0.8, 0.75)$. (d). Response $x_1(t)$ for $(\alpha, a, H) = (2.8, 0.8, 0.75)$.

## 3.3 RESPONSES OF CLASS II FRACTIONAL VIBRATORS DRIVEN BY GFGN

### 3.3.1 Computation Methods

For a class II fractional vibrator, its motion equation is given by

$$m\frac{d^2x_2(t)}{dt^2}+c\frac{d^\beta x_2(t)}{dt^\beta}+k\frac{dx_2(t)}{dt}=f(t). \tag{3.9}$$

In (3.9), $0<\beta<2$, $x_2(t)$ is the response of a class II fractional vibrator and $c$ is the primary damping.

**Theorem 3.3 (PSD response II)**

Let $S_{xx2}(\omega)$ be the PSD of $x_2(t)$. Then,

$$S_{xx2}(\omega)=\frac{-V_H H(2H-1)\sin\left[a\pi(H-1)\right]\Gamma(2aH-2a+1)|\omega|^{-2a(H-1)-1}}{k^2\left\{\left[1-\gamma^2\left(1-\frac{c}{m}\omega^{\beta-2}\cos\frac{\beta\pi}{2}\right)\right]^2+\left(\frac{2\varsigma\omega^\beta}{\omega_n}\sin\frac{\beta\pi}{2}\right)^2\right\}}. \tag{3.10}$$

In (3.10), $\gamma=\frac{\omega}{\omega_n}$.

*Proof.* Because $S_{xx2}(\omega)=S_{ff}(\omega)|H_2(\omega)|^2$, where $H_2(\omega)$ is given by (Chapter 7 in Volume I, Li [1–3]),

$$H_2(\omega)=\frac{1/k}{1-\gamma^2\left(1-\frac{c}{m}\omega^{\beta-2}\cos\frac{\beta\pi}{2}\right)+i\frac{2\varsigma\omega^\beta\sin\frac{\beta\pi}{2}}{\omega_n}}, \tag{3.11}$$

(3.10) is true. This finishes the proof.

**Theorem 3.4 (cross-PSD response II)**

Denote by $S_{fx2}(\omega)$ the cross PSD between $f(t)$ and $x_2(t)$. Then,

$$S_{fx2}(\omega)=\frac{-V_H H(2H-1)\sin\left[a\pi(H-1)\right]\Gamma(2aH-2a+1)|\omega|^{-2a(H-1)-1}}{k\left[1-\gamma^2\left(1-2\varsigma\omega_n\omega^{\beta-2}\cos\frac{\beta\pi}{2}\right)+i\frac{2\varsigma\omega^\beta}{\omega_n}\sin\frac{\beta\pi}{2}\right]}. \tag{3.12}$$

*Proof.* Since $S_{fx2}(\omega) = S_{ff}(\omega)H_2(\omega)$ and (3.11), (3.12) holds. The proof is finished.

Let $r_{xx2}(\tau)$ be the ACF of $x_2(t)$. Denote by $h_2(t)$ the impulse response of a class II fractional vibrator. Then, $r_{xx2}(\tau) \leftrightarrow S_{xx2}(\omega)$. According to the Wiener-Khinchin relation and the Wiener-Lee relation, we have

$$r_{xx2}(\tau) = r_{ff}(\tau) * h_2(\tau) * h_2(-\tau). \tag{3.13}$$

Let $r_{fx2}(\tau)$ be the cross-correlation between $f(t)$ and $x_2(t)$. Then, $r_{fx2}(\tau) \leftrightarrow S_{fx2}(\omega)$ and

$$r_{fx2}(\tau) = r_{ff}(\tau) * h_2(\tau). \tag{3.14}$$

In (3.14) and (3.14) (Chapter 7 in Volume I, Li [1–3]), $h_2(t)$ is expressed by

$$h_2(t) = \frac{e^{-\frac{\varsigma\omega_n\omega^{\beta-1}\sin\frac{\beta\pi}{2}}{1-\frac{c}{m}\omega^{\beta-2}\cos\frac{\beta\pi}{2}}t} \sin\dfrac{\omega_n\sqrt{1-\dfrac{\varsigma^2\omega^{2(\beta-1)}\sin^2\frac{\beta\pi}{2}}{1-\frac{c}{m}\omega^{\beta-2}\cos\frac{\beta\pi}{2}}}}{\sqrt{1-\frac{c}{m}\omega^{\beta-2}\cos\frac{\beta\pi}{2}}}t}{\omega_n m\sqrt{1-\frac{c}{m}\omega^{\beta-2}\cos\frac{\beta\pi}{2}}\sqrt{1-\dfrac{\varsigma^2\omega^{2(\beta-1)}\sin^2\frac{\beta\pi}{2}}{1-\frac{c}{m}\omega^{\beta-2}\cos\frac{\beta\pi}{2}}}} u(t). \tag{3.15}$$

## 3.3.2 Effect of $\beta$ on Responses

Some plots of $|H_2(\omega)|^2$ are indicated in Figure 3.7. Figure 3.8 shows some plots of response PSD $S_{xx2}(\omega)$. Figure 3.9 shows a resonance curve when $\beta = 1$. Figure 3.10 illustrates some plots of $|S_{fx2}(\omega)|$. Figure 3.11 exhibits the effect of $\beta$ on the fluctuation range of $x_2(t)$. From Figures 3.9 to 3.11, we see that there is effect of $\beta$ on the responses to class II fractional vibration systems driven by gfGn.

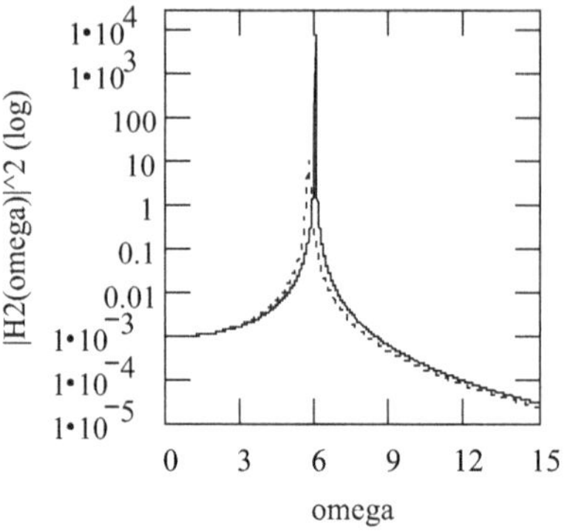


FIGURE 3.7 Plots of $|H_2(\omega)|^2$ (log) with $\beta = 0.1$ (solid), 1.9 (dot), when $m = 1$, $c = 0.1$, and $k = 36$.

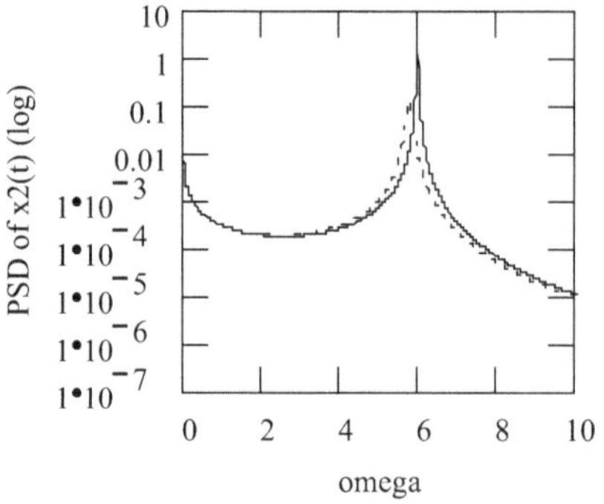


FIGURE 3.8 Plots of $S_{xx2}(\omega)$ (log) with $\beta = 0.4$ (solid), 1.8 (dot), when $m = 1$, $c = 0.1$, and $k = 36$ when $(a, H) = (0.8, 0.95)$.

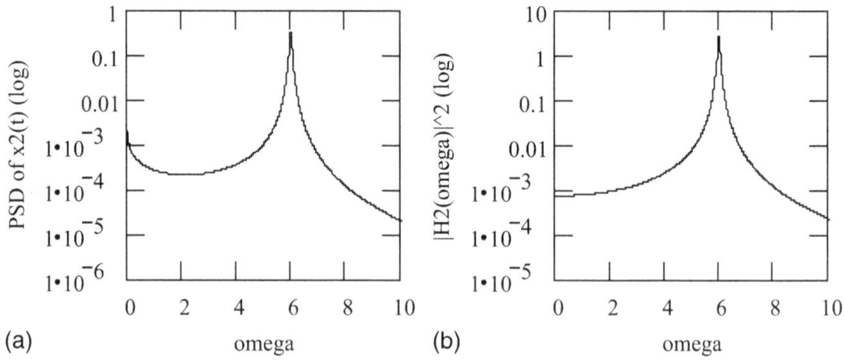


FIGURE 3.9 Resonance when $\beta = 1$, $m = 1$, $c = 0.1$, $k = 36$. (a). Plot of $|H_2(\omega)|^2$ (log). (b). Plot of response PSD $S_{xx2}(\omega)$ (log) when $(a, H) = (0.8, 0.75)$.

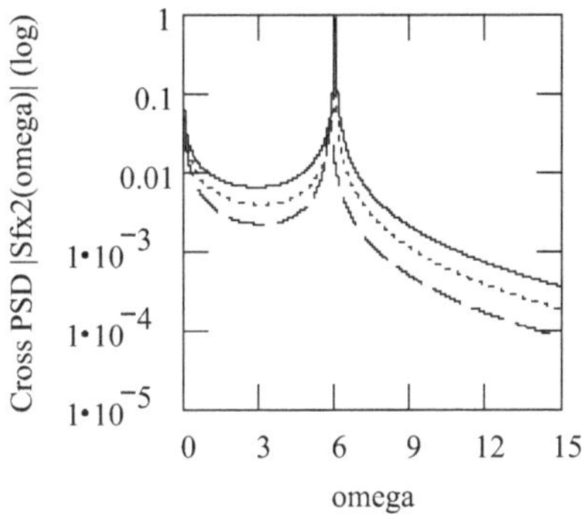


FIGURE 3.10 Plots of $|S_{fx2}(\omega)|$ (log) when $m = 1$, $c = 0.1$, and $k = 36$ with ($\beta$, $a$, $H$) = (0.4, 0.8, 0.75) (solid), ($\beta$, $a$, $H$) = (1, 0.6, 0.75) (dot), ($\beta$, $a$, $H$) = (1.8, 0.4, 0.75) (dash).

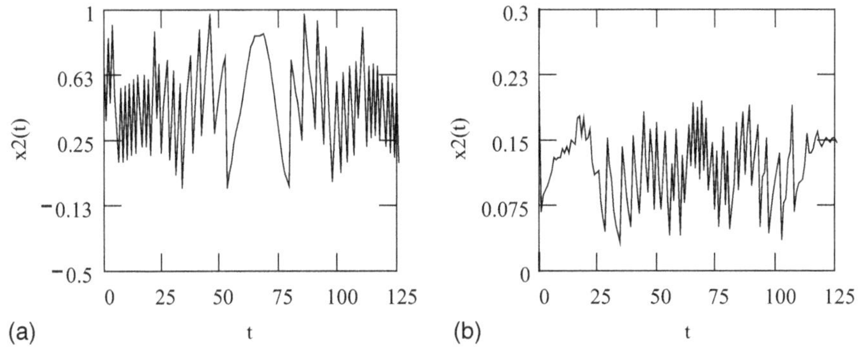


FIGURE 3.11 Response $x_2(t)$ when $m = 1$, $c = 0.1$, $k = 36$, and ($a$, $H$) = (0.8, 0.75). (a). $x_2(t)$ for $\beta = 0.4$. (b). $x_2(t)$ for $\beta = 1.4$.

## 3.4 RESPONSES OF CLASS III FRACTIONAL VIBRATORS DRIVEN BY GFGN

### 3.4.1 Computations

The motion equation of a class III fractional vibrator is given by

$$m\frac{d^{\alpha}x_3(t)}{dt^{\alpha}}+c\frac{d^{\beta}x_3(t)}{dt^{\beta}}+kx_3(t)=f(t). \tag{3.16}$$

In (3.16), $x_3(t)$ is the response of a class III fractional vibrator.

**Theorem 3.5 (PSD response III)**

Let $S_{xx3}(\omega)$ be the PSD of $x_3(t)$. Then,

$$S_{xx3}(\omega)=\frac{-V_H H(2H-1)\sin\left[a\pi(H-1)\right]\Gamma(2aH-2a+1)\left|\omega\right|^{-2a(H-1)-1}}{k^2\left\{\left[1-\gamma^2\left(\omega^{\alpha-2}\left|\cos\frac{\alpha\pi}{2}\right|-2\varsigma\omega_n\omega^{\beta-2}\cos\frac{\beta\pi}{2}\right)\right]^2+\left[\frac{\gamma\left(\omega^{\alpha-1}\sin\frac{\alpha\pi}{2}+2\varsigma\omega_n\omega^{\beta-1}\sin\frac{\beta\pi}{2}\right)}{\omega_n\left(\omega^{\alpha-2}\left|\cos\frac{\alpha\pi}{2}\right|-2\varsigma\omega_n\omega^{\beta-2}\cos\frac{\beta\pi}{2}\right)}\right]^2\right\}}. \tag{3.17}$$

*Proof.* As $S_{xx3}(\omega) = S_{ff}(\omega)|H_3(\omega)|^2$, where $H_3(\omega)$ is expressed by (Chapter 7 in Volume I or Li [1–3])

$$H_3(\omega)=\frac{1/k}{1-\gamma^2\left(\omega^{\alpha-2}\left|\cos\frac{\alpha\pi}{2}\right|-2\varsigma\omega_n\omega^{\beta-2}\cos\frac{\beta\pi}{2}\right)+i\frac{\gamma\left(\omega^{\alpha-1}\sin\frac{\alpha\pi}{2}+2\varsigma\omega_n\omega^{\beta-1}\sin\frac{\beta\pi}{2}\right)}{\omega_n\left(\omega^{\alpha-2}\left|\cos\frac{\alpha\pi}{2}\right|-2\varsigma\omega_n\omega^{\beta-2}\cos\frac{\beta\pi}{2}\right)}}, \tag{3.18}$$

(3.17) holds. The proof is finished.

**Theorem 3.6 (cross-PSD response III)**

Let $S_{fx3}(\omega)$ be the cross PSD between $f(t)$ and $x_3(t)$. Then,

$$S_{fx3}(\omega)=\frac{-k^{-1}V_H H(2H-1)\sin\left[a\pi(H-1)\right]\Gamma(2aH-2a+1)\left|\omega\right|^{-2a(H-1)-1}}{1-\gamma^2\left(\omega^{\alpha-2}\left|\cos\frac{\alpha\pi}{2}\right|-2\varsigma\omega_n\omega^{\beta-2}\cos\frac{\beta\pi}{2}\right)+i\frac{\gamma\left(\omega^{\alpha-1}\sin\frac{\alpha\pi}{2}+2\varsigma\omega_n\omega^{\beta-1}\sin\frac{\beta\pi}{2}\right)}{\omega_n\left(\omega^{\alpha-2}\left|\cos\frac{\alpha\pi}{2}\right|-2\varsigma\omega_n\omega^{\beta-2}\cos\frac{\beta\pi}{2}\right)}}. \tag{3.19}$$

*Proof.* Performing the operation of $S_{fx3}(\omega) = S_{ff}(\omega)H_3(\omega)$ and taking into account (3.18) yields (3.19). The proof ends.

Let $F^{-1}$ be the operator of the inverse Fourier transform. Denote by $r_{xx3}(\tau) = F^{-1}[S_{xx3}(\omega)]$ the ACF of $x_3(t)$. Denote by $h_3(\tau)$ the impulse response of a class III fractional vibrator. According to the Wiener-Khinchin relation and the Wiener-Lee relation, the ACF response $r_{xx3}(\tau)$ is expressed by

$$r_{xx3}(\tau) = r_{ff}(\tau) * h_3(\tau) * h_3(-\tau). \tag{3.20}$$

Let $r_{fx3}(\tau) = F^{-1}[S_{fx3}(\omega)]$ be the cross-correlation between $f(t)$ and $x_3(t)$. Then, the cross-correlation response $r_{fx3}(\tau)$ is given by

$$r_{fx3}(\tau) = r_{ff}(\tau) * h_3(\tau). \tag{3.21}$$

In (3.20) and (3.21) (Chapter 7 in Volume I, Li [1–3]),

$$h_3(t) = \frac{e^{-\frac{m\omega^{\alpha-1}\sin\frac{\alpha\pi}{2}+c\omega^{\beta-1}\sin\frac{\beta\pi}{2}}{2\sqrt{-\left(m\omega^{\alpha-2}\cos\frac{\alpha\pi}{2}+c\omega^{\beta-2}\cos\frac{\beta\pi}{2}\right)k}}\omega_{\mathrm{eqn}3}t}\sin\omega_{\mathrm{eqd}3}t}{-\left(m\omega^{\alpha-2}\cos\frac{\alpha\pi}{2}+c\omega^{\beta-2}\cos\frac{\beta\pi}{2}\right)\omega_{\mathrm{eqd}3}}u(t). \tag{3.22}$$

Refer to Chapter 7 in Volume I or Li [1–3] for the expressions of $\omega_{\mathrm{eqn3}}$ and $\omega_{\mathrm{eqd3}}$ in (3.22).

### 3.4.2 Effect of $(\alpha, \beta)$ on Responses

Some plots of $|H_3(\omega)|^2$ are shown in Figure 3.12. Figure 3.13 indicates some plots of response PSD $S_{xx3}(\omega)$. Some plots of $|S_{fx3}(\omega)|$ are demonstrated in Figure 3.14. Figures 3.13 and 3.14 show that the effect of $(\alpha, \beta)$ on the responses to class III fractional vibrators driven by gfGn is significant.

## 3.5 RESPONSES OF CLASS IV FRACTIONAL VIBRATORS DRIVEN BY GFGN

### 3.5.1 Computations

The motion equation of a class IV fractional vibrator is in the form

$$m\frac{d^{\alpha}x_4(t)}{dt^{\alpha}} + k\frac{d^{\lambda}x_4(t)}{dt^{\lambda}} = f(t). \tag{3.23}$$

In (3.23), $1 \le \lambda < 1$, $x_4(t)$ is the response of a class IV fractional vibrator.

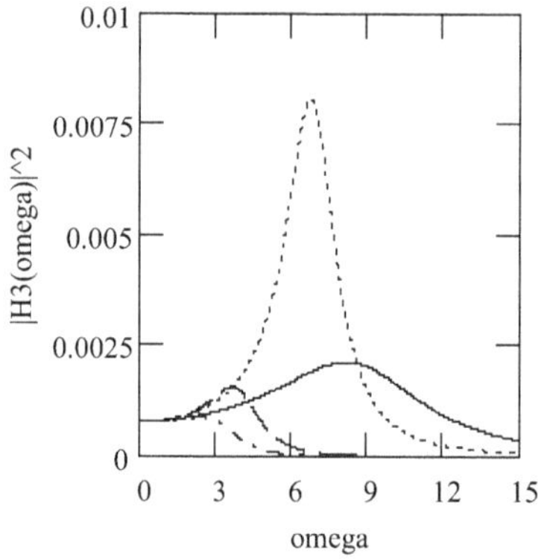


FIGURE 3.12 Plots of $|H_3(\omega)|^2$ when $m = 1$, $c = 0.1$, and $k = 36$, for $(\alpha, \beta) = (1.6, 0.8)$ (solid), (1.8, 1.8) (dot), (2.5, 0.8) (dash), (2.8, 1.8) (dash dot).

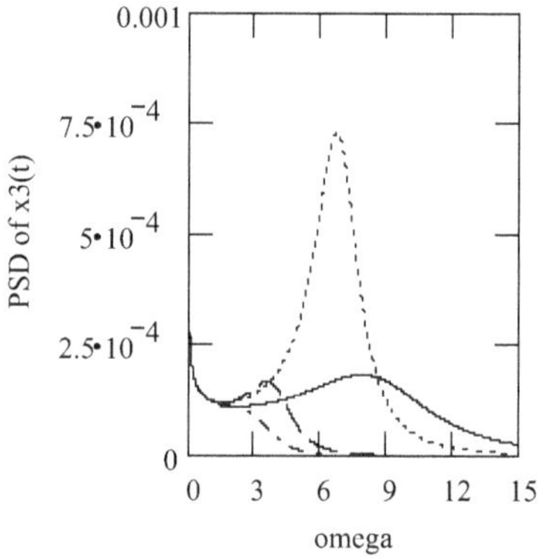


FIGURE 3.13 Plots of $S_{xx3}(\omega)$ when $m = 1$, $c = 0.1$, $k = 36$, and $(a, H) = (0.8, 0.55)$ for $(\alpha, \beta) = (1.6, 0.8)$ (solid), (1.8, 1.8) (dot), (2.5, 0.8) (dash), (2.8, 1.8) (dash dot).

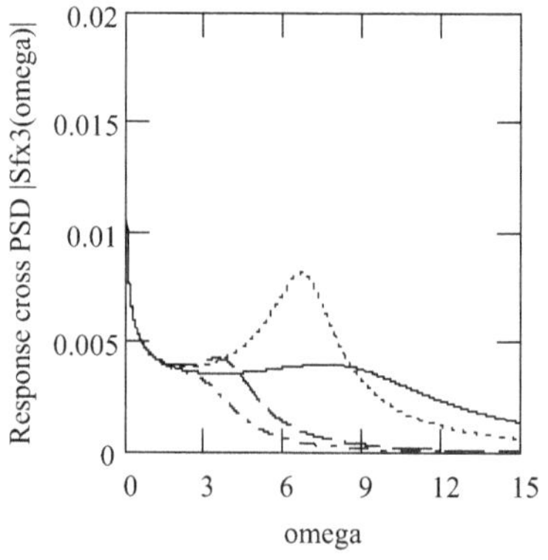


FIGURE 3.14 Plots of $|S_{fx3}(\omega)|$ when $m = 1$, $c = 0.1$, $k = 36$, and $(a, H) = (0.8, 0.55)$ for $(\alpha, \beta) = (1.6, 0.8)$ (solid), (1.8, 1.8) (dot), (2.5, 0.8) (dash), (2.8, 1.8) (dash dot).

**Theorem 3.7 (PSD response IV)**

Let $S_{xx4}(\omega)$ be the PSD of $x_4(t)$. Then,

$$S_{xx4}(\omega)=\frac{-V_H H(2H-1)\sin\left[a\pi(H-1)\right]\Gamma(2aH-2a+1)\left|\omega\right|^{-2a(H-1)-1}}{k^2\omega^{2\lambda}\cos^2\frac{\lambda\pi}{2}\left[\left(1-\gamma^2\frac{-\omega^{\alpha-2}\cos\frac{\alpha\pi}{2}}{\omega^{\lambda}\cos\frac{\lambda\pi}{2}}\right)^2+4\left(\gamma\frac{m\omega^{\alpha-1}\sin\frac{\alpha\pi}{2}+k\omega^{\lambda-1}\sin\frac{\lambda\pi}{2}}{2\sqrt{mk\omega^{\alpha+\lambda-2}\left|\cos\frac{\alpha\pi}{2}\right|\left|\cos\frac{\lambda\pi}{2}\right|}}\sqrt{\frac{-\omega^{\alpha-2}\cos\frac{\alpha\pi}{2}}{\omega^{\lambda}\cos\frac{\lambda\pi}{2}}}\right)^2\right]}. \tag{3.24}$$

*Proof.* Doing the operation of $S_{xx4}(\omega) = S_{ff}(\omega)|H_4(\omega)|^2$ and following Chapter 7 in Volume I or Li [1] or [2] about $H_4(\omega)$ in the form

$$H_4(\omega)=\frac{1}{k\omega^{\lambda}\cos\frac{\lambda\pi}{2}\left(1-\gamma^2\frac{-\omega^{\alpha-2}\cos\frac{\alpha\pi}{2}}{\omega^{\lambda}\cos\frac{\lambda\pi}{2}}+i2\gamma\frac{m\omega^{\alpha-1}\sin\frac{\alpha\pi}{2}+k\omega^{\lambda-1}\sin\frac{\lambda\pi}{2}}{2\sqrt{mk\omega^{\alpha+\lambda-2}\left|\cos\frac{\alpha\pi}{2}\right|\left|\cos\frac{\lambda\pi}{2}\right|}}\sqrt{\frac{-\omega^{\alpha-2}\cos\frac{\alpha\pi}{2}}{\omega^{\lambda}\cos\frac{\lambda\pi}{2}}}\right)} \tag{3.25}$$

yields (3.24). The proof is finished.

**Theorem 3.8 (cross-PSD response IV)**

Denote by $S_{fx4}(\omega)$ the cross PSD between $f(t)$ and $x_4(t)$. Then,

$$S_{fx4}(\omega)=\frac{-V_H H(2H-1)\sin\left[a\pi(H-1)\right]\Gamma(2aH-2a+1)\left|\omega\right|^{-2a(H-1)-1}}{k\omega^{\lambda}\cos\frac{\lambda\pi}{2}\left(1-\gamma^2\frac{-\omega^{\alpha-2}\cos\frac{\alpha\pi}{2}}{\omega^{\lambda}\cos\frac{\lambda\pi}{2}}+i2\gamma\frac{m\omega^{\alpha-1}\sin\frac{\alpha\pi}{2}+k\omega^{\lambda-1}\sin\frac{\lambda\pi}{2}}{2\sqrt{mk\omega^{\alpha+\lambda-2}\left|\cos\frac{\alpha\pi}{2}\right|\left|\cos\frac{\lambda\pi}{2}\right|}}\sqrt{\frac{-\omega^{\alpha-2}\cos\frac{\alpha\pi}{2}}{\omega^{\lambda}\cos\frac{\lambda\pi}{2}}}\right)}. \tag{3.26}$$

*Proof.* Owing to $S_{fx4}(\omega) = S_{ff}(\omega)H_4(\omega)$ and (3.25) about $H_4(\omega)$, we have (3.26) (Li [13]). This finishes the proof.

Let $r_{xx4}(\tau)$ be the ACF of $x_4(t)$. Denote by $h_4(\tau)$ the impulse response of a class IV fractional vibrator. Let $r_{fx4}(\tau)$ be the cross-correlation between $f(t)$ and $x_4(t)$. Then, the ACF response $r_{xx4}(\tau)$ is in the form

$$r_{xx4}(\tau) = r_{ff}(\tau) * h_4(\tau) * h_4(-\tau), \tag{3.27}$$

and the cross-correlation response $r_{fx4}(\tau)$ is given by

$$r_{fx4}(\tau) = r_{ff}(\tau) * h_4(\tau). \tag{3.28}$$

In (3.27) and (3.28), $h_4(t)$ is given by (Chapter 7 in Volume I, Li [1] or [2])

$$h_4(t) = e^{-\frac{m\omega^{\alpha-1}\sin\frac{\alpha\pi}{2}+k\omega^{\lambda-1}\sin\frac{\lambda\pi}{2}}{2\sqrt{mk\omega^{\alpha+\lambda-2}\left|\cos\frac{\alpha\pi}{2}\right|\cos\frac{\lambda\pi}{2}}}\sqrt{\frac{\omega^{\lambda}\cos\frac{\lambda\pi}{2}}{-\omega^{\alpha-2}\cos\frac{\alpha\pi}{2}}}\,\omega_n t}\ \frac{1}{m_{eq4}\omega_{eqd4}}\sin\omega_{eqd4}t\,u(t). \tag{3.29}$$

Refer to Chapter 7 in Volume I or Li [1] or [2] for the expressions of $m_{eq4}$, $\omega_{eqn4}$, and $\omega_{eqd4}$ in (3.29).

## 3.5.2 Effect of ($\alpha$, $\lambda$) on Responses

We illustrate some plots of $|H_4(\omega)|^2$ in Figure 3.15. Figure 3.16 indicates some plots of response PSD $S_{xx4}(\omega)$. Some plots of $|S_{fx4}(\omega)|$ are shown in

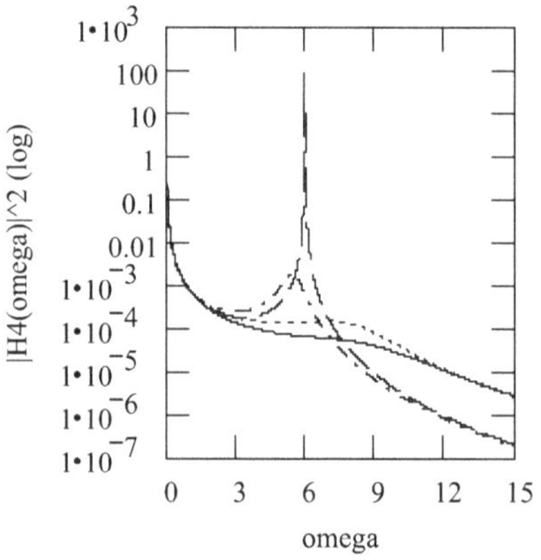


FIGURE 3.15 Plots of $|H_4(\omega)|^2$ in log when $m = 1$, $c = 0$, and $k = 36$, for $(\alpha, \lambda) = (2.5, 0.9)$ (solid), $(2.5, 0.8)$ (dot), $(2.9, 0.9)$ (dash), $(2.9, 0.8)$ (dash dot).

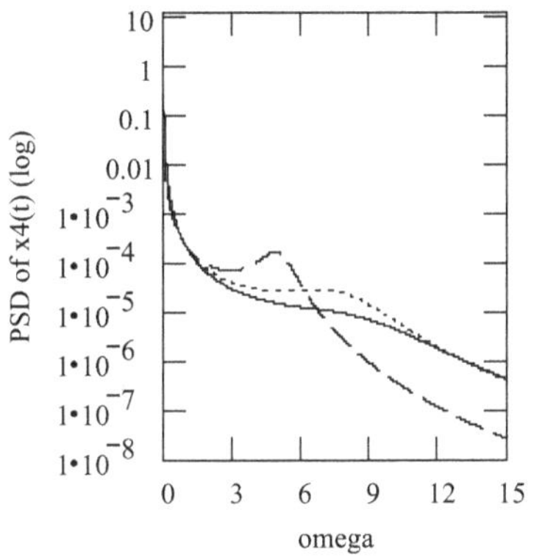


FIGURE 3.16 Plots of $S_{xx4}(\omega)$ in log when $m = 1$, $c = 0$, $k = 36$, and $(a, H) = (0.9, 0.55)$ for $(\alpha, \lambda) = (2.5, 0.9)$ (solid), $(2.5, 0.8)$ (dot), $(2.9, 0.7)$ (dash), $(2.9, 0.6)$ (dash dot).

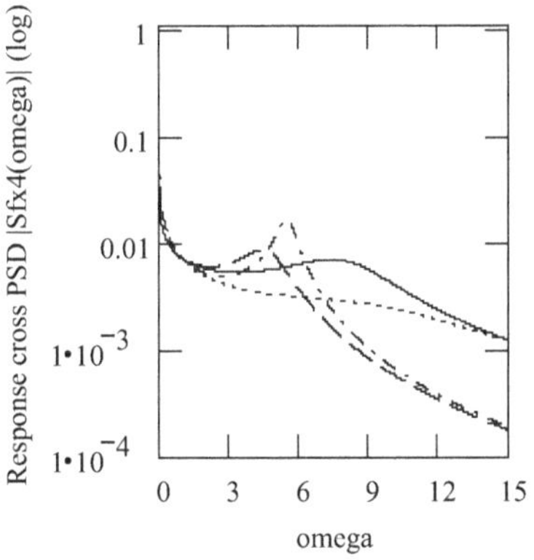


FIGURE 3.17 Plots of $|S_{fx4}(\omega)|$ when $m = 1$, $c = 0$, $k = 36$, and $(a, H) = (0.9, 0.55)$ for $(\alpha, \lambda) = (1.9, 0.2)$ (solid), $(1.9, 0.4)$ (dot), $(2.5, 0.2)$ (dash), $(2.5, 0.4)$ (dash dot).

Figure 3.17. Figures 3.17 and 3.18 exhibit that the effect of $(\alpha, \lambda)$ on the responses to class IV fractional vibrators driven by gfGn is noticeable.

## 3.6 RESPONSES OF CLASS V FRACTIONAL VIBRATORS DRIVEN BY GFGN

### 3.6.1 Computation Methods

The motion equation of a class V fractional vibrator is expressed by

$$m\frac{d^2 x_5(t)}{dt^2} + k\frac{d^\lambda x_5(t)}{dt^\lambda} = f(t). \tag{3.30}$$

In (3.30), $x_5(t)$ is the response of a class V fractional vibrator.

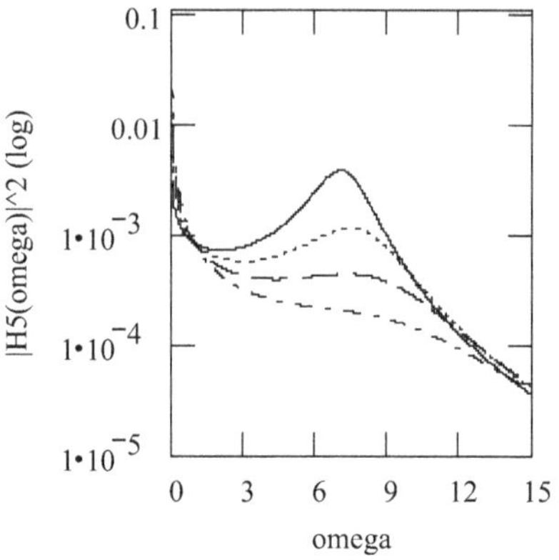


FIGURE 3.18 Plots of $|H_5(\omega)|^2$ (log) when $m = 1$, $c = 0$, and $k = 36$, for $\lambda = 0.2$ (solid), 0.3 (dot), 0.4 (dash), 0.5 (dash dot).

**Theorem 3.9 (PSD response V)**

Denote by $S_{xx5}(\omega)$ the PSD of $x_5(t)$. Then,

$$S_{xx5}(\omega)=\frac{-V_H H(2H-1)\sin\left[a\pi(H-1)\right]\Gamma(2aH-2a+1)\left|\omega\right|^{-2a(H-1)-1}}{k^2\omega^{2\lambda}\cos^2\frac{\lambda\pi}{2}\left[\left(1-\frac{\gamma^2}{\omega^\lambda\cos\frac{\lambda\pi}{2}}\right)^2+4\gamma^2\left(\frac{k\omega^{\lambda-1}\sin\frac{\lambda\pi}{2}}{2\sqrt{mk\omega^\lambda\cos\frac{\lambda\pi}{2}}}\sqrt{\frac{1}{\omega^\lambda\cos\frac{\lambda\pi}{2}}}\right)^2\right]}. \tag{3.31}$$

*Proof.* Performing the operation of $S_{xx5}(\omega) = S_{ff}(\omega)|H_5(\omega)|^2$, where $H_5(\omega)$ is given by (Chapter 7 in Volume I, or Li [1, 2]),

$$H_5(\omega)=\frac{1}{k\omega^\lambda\cos\frac{\lambda\pi}{2}\left(1-\frac{\gamma^2}{\omega^\lambda\cos\frac{\lambda\pi}{2}}+i2\gamma\frac{k\omega^{\lambda-1}\sin\frac{\lambda\pi}{2}}{2\sqrt{mk\omega^\lambda\cos\frac{\lambda\pi}{2}}}\sqrt{\frac{1}{\omega^\lambda\cos\frac{\lambda\pi}{2}}}\right)}, \tag{3.32}$$

(3.31) holds. The proof is finished.

**Theorem 3.10 (cross-PSD response V)**

Let $S_{fx5}(\omega)$ be the cross-PSD between $f(t)$ and $x_5(t)$. Then,

$$S_{fx5}(\omega)=\frac{-V_H H(2H-1)\sin\left[a\pi(H-1)\right]\Gamma(2aH-2a+1)\left|\omega\right|^{-2a(H-1)-1}}{k\omega^{\lambda}\cos\frac{\lambda\pi}{2}\left(1-\frac{\gamma^2}{\omega^{\lambda}\cos\frac{\lambda\pi}{2}}+i2\gamma\frac{k\omega^{\lambda-1}\sin\frac{\lambda\pi}{2}}{2\sqrt{mk\omega^{\lambda}\cos\frac{\lambda\pi}{2}}}\sqrt{\frac{1}{\omega^{\lambda}\cos\frac{\lambda\pi}{2}}}\right)}. \tag{3.33}$$

*Proof.* Doing the operation of $S_{fx5}(\omega) = S_{ff}(\omega)H_5(\omega)$ and considering (3.32) yields (3.33). This finishes the proof.

Let $r_{xx5}(\tau)$ be the ACF of $x_5(t)$. Denote by $h_5(\tau)$ the impulse response of a class V fractional vibrator. Then, one has the ACF response expressed by

$$r_{xx5}(\tau) = r_{ff}(\tau)*h_5(\tau)*h_5(-\tau). \tag{3.34}$$

Let $r_{fx5}(\tau)$ be the cross-correlation between $f(t)$ and $x_5(t)$. Then, the cross-correlation response is given by

$$r_{fx5}(\tau) = r_{ff}(\tau)*h_5(\tau). \tag{3.35}$$

In (3.34) and (3.35) (Chapter 7 in Volume I, or Li [1, 2]),

$$h_5(t)=e^{-\frac{k\omega^{\lambda-1}\sin\frac{\lambda\pi}{2}}{2\sqrt{mk\omega^{\lambda}\cos\frac{\lambda\pi}{2}}}\sqrt{\omega^{\lambda}\cos\frac{\lambda\pi}{2}}\,\omega_n t}\frac{1}{m\omega_{\text{eqd5}}}\sin\omega_{\text{eqd5}}tu(t). \tag{3.36}$$

Refer to Chapter 7 in Volume I or Li [1] or Li [2] for the expression of $\omega_{\text{eqd5}}$ in (3.36).

### 3.6.2 Effect of $\lambda$ on Responses

Figure 3.18 illustrates some plots of $|H_5(\omega)|^2$. Figure 3.19 indicates some plots of response PSD $S_{xx5}(\omega)$ and Figure 3.20 shows some plots of $|S_{fx5}(\omega)|$. Figures 3.19 and 3.20 indicate that the effect of $\lambda$ on the responses to class V fractional vibration systems driven by gfGn is considerable.

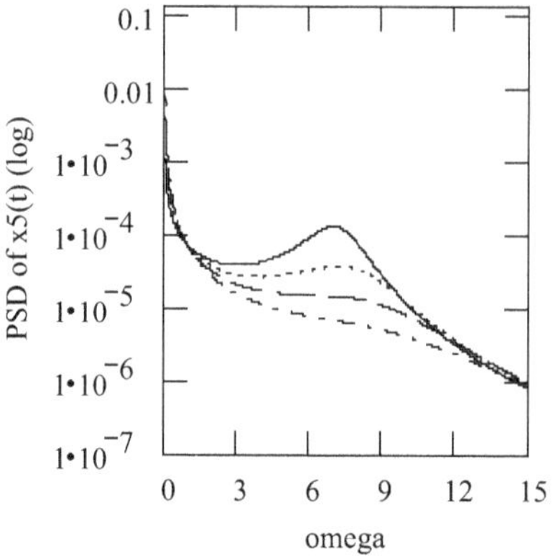


FIGURE 3.19 Plots of $S_{xx5}(\omega)$ in log when $m = 1$, $c = 0$, $k = 36$, and $(a, H) = (0.6, 0.55)$ for $\lambda = 0.2$ (solid), 0.4 (dot), 0.6 (dash), 0.8 (dash dot).

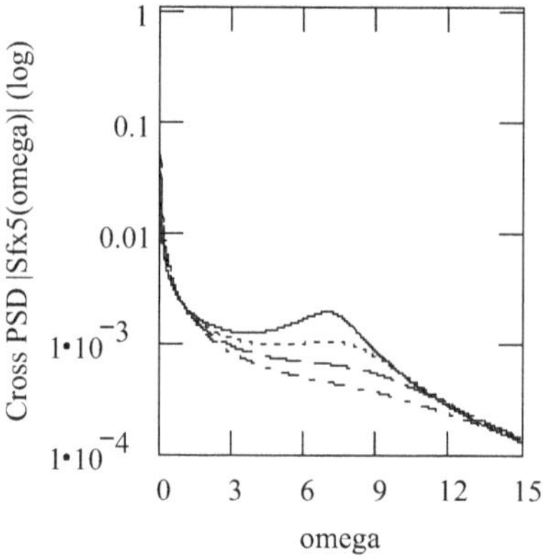


FIGURE 3.20 Plots of $|S_{fx5}(\omega)|$ in log when $m = 1$, $c = 0$, $k = 36$ for $\lambda = 0.2$ (solid), 0.3 (dot), 0.4 (dash), 0.5 (dash dot).

## 3.7 RESPONSES OF CLASS VI FRACTIONAL VIBRATORS DRIVEN BY GFGN

### 3.7.1 Computations

The motion equation of a class VI fractional vibrator is in the form

$$m\frac{d^{\alpha}x_6(t)}{dt^{\alpha}}+c\frac{d^{\beta}x_6(t)}{dt^{\beta}}+k\frac{d^{\lambda}x_6(t)}{dt^{\lambda}}=f(t),\quad 1<\alpha<3,\ 0<\beta<2,\ 0\leq\lambda<1, \tag{3.37}$$

In (3.37), $x_6(t)$ is the response of a class VI fractional vibrator.

**Theorem 3.11 (PSD response VI)**

Let $S_{xx6}(\omega)$ be the PSD of $x_6(t)$. Then,

$$S_{xx6}(\omega)=\frac{1}{k^2}\frac{-V_H H(2H-1)\sin\left[a\pi(H-1)\right]\Gamma(2aH-2a+1)\left|\omega\right|^{-2a(H-1)-1}}{\left[\omega^{\lambda}\cos\frac{\lambda\pi}{2}+\gamma^2\left(\omega^{\alpha-2}\cos\frac{\alpha\pi}{2}+2\varsigma\omega_n\omega^{\beta-2}\cos\frac{\beta\pi}{2}\right)\right]^2+\gamma^2\left(\omega^{\alpha-1}\sin\frac{\alpha\pi}{2}+2\zeta\omega_n\omega^{\beta-1}\sin\frac{\beta\pi}{2}+\omega_n^2\omega^{\lambda-1}\sin\frac{\lambda\pi}{2}\right)^2}. \tag{3.38}$$

*Proof.* According to Chapter 7 in Volume I or Li [1, 2], $H_6(\omega)$ is given by

$$H_6(\omega)=\frac{1}{k\left[\omega^{\lambda}\cos\frac{\lambda\pi}{2}+\gamma^2\left(\omega^{\alpha-2}\cos\frac{\alpha\pi}{2}+2\varsigma\omega_n\omega^{\beta-2}\cos\frac{\beta\pi}{2}\right)+i\gamma\left(\omega^{\alpha-1}\sin\frac{\alpha\pi}{2}+2\zeta\omega_n\omega^{\beta-1}\sin\frac{\beta\pi}{2}+\omega_n^2\omega^{\lambda-1}\sin\frac{\lambda\pi}{2}\right)\right]}. \tag{3.39}$$

Taking into account $S_{xx6}(\omega) = S_{ff}(\omega)|H_6(\omega)|^2$, therefore, (3.38) holds. The proof ends.

**Theorem 3.12 (cross-PSD response VI)**

Denote by $S_{fx6}(\omega)$ the cross-PSD between $f(t)$ and $x_6(t)$. Then,

$$S_{fx6}(\omega)=\frac{-V_H H(2H-1)\sin\left[a\pi(H-1)\right]\Gamma(2aH-2a+1)\left|\omega\right|^{-2a(H-1)-1}}{k\left[\omega^{\lambda}\cos\frac{\lambda\pi}{2}+\gamma^2\left(\omega^{\alpha-2}\cos\frac{\alpha\pi}{2}+2\varsigma\omega_n\omega^{\beta-2}\cos\frac{\beta\pi}{2}\right)+i\gamma\left(\omega^{\alpha-1}\sin\frac{\alpha\pi}{2}+2\zeta\omega_n\omega^{\beta-1}\sin\frac{\beta\pi}{2}+\omega_n^2\omega^{\lambda-1}\sin\frac{\lambda\pi}{2}\right)\right]}. \tag{3.40}$$

*Proof.* Using $S_{fx6}(\omega)=S_{ff}(\omega)H_6(\omega)$ and (3.39) yields (3.40). The proof is finished.

In the time domain, the ACF response is

$$r_{xx6}(\tau) = r_{ff}(\tau)*h_6(\tau)*h_6(-\tau), \tag{3.41}$$

where $r_{xx6}(\tau)$ is the ACF of $x_6(t)$ and $h_6(\tau)$ is the impulse response of a class VI fractional vibrator. Let $r_{fx6}(\tau)$ be the cross-correlation between $f(t)$ and $x_6(t)$. Then, the cross-correlation response $r_{fx6}(\tau)$ is given by

$$r_{fx6}(\tau) = r_{ff}(\tau) * h_6(\tau). \tag{3.42}$$

In (3.41) and (3.42),

$$h_6(t) = e^{-\zeta_{eq6}\omega_{eqn6}t} \frac{1}{m_{eq6}\omega_{eqd6}} \sin \omega_{eqd6} t u(t). \tag{3.43}$$

Refer to Chapter 7 in Volume I or Li [1] or Li [2] for the expressions of $\zeta_{eq6}$, $m_{eq6}$, $\omega_{eqn6}$, and $\omega_{eqd6}$ in (3.43).

### 3.7.2 Effect of $(\alpha, \beta, \lambda)$ on Responses

Figure 3.21 shows some plots of $|H_6(\omega)|^2$. Figure 3.22 is used for some plots of $S_{xx6}(\omega)$ and Figure 3.23 for some plots of $|S_{fx6}(\omega)|$. As can be seen from Figures 3.22 and 3.23, the effect of $(\alpha, \beta, \lambda)$ on the responses to class VI fractional vibrators under the excitation of gfGn is significant.

## 3.8 RESPONSES OF CLASS VII FRACTIONAL VIBRATORS DRIVEN BY GFGN

### 3.8.1 Computations

The motion equation of a class VII fractional vibrator is expressed by

$$m\frac{d^2x_7(t)}{dt^2} + c\frac{d^\beta x_7(t)}{dt^\beta} + k\frac{d^\lambda x_7(t)}{dt^\lambda} = f(t), \quad 0 < \beta < 2,\ 0 \le \lambda < 1, \tag{3.44}$$

In (3.44), $x_7(t)$ is the response of a class VII fractional vibrator.

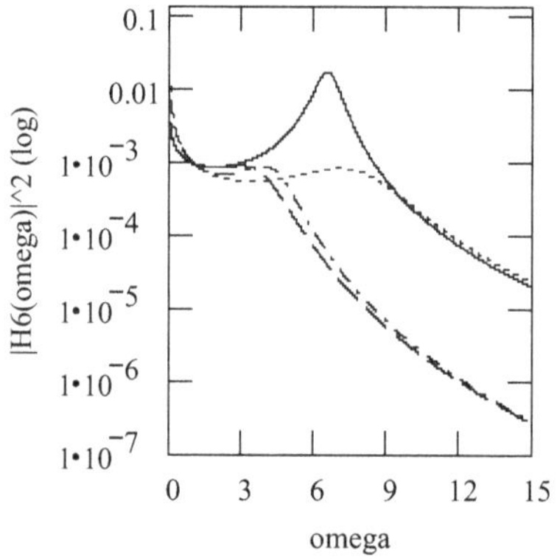


FIGURE 3.21 Plots of $|H_6(\omega)|^2$ (log) when $m = 1$, $c = 0.1$, and $k = 36$, for $(\alpha, \beta, \lambda) = (2.1, 0.8, 0.2)$ (solid), (2.1, 1.8, 0.4) (dot), (2.8, 0.4, 0.2) (dash), (2.8, 0.8, 0.4) (dash dot).

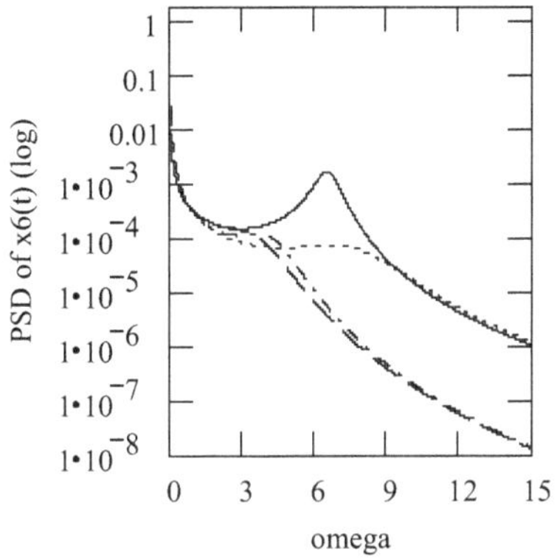


FIGURE 3.22 Plots of response PSD $S_{xx6}(\omega)$ in log when $m = 1$, $c = 0.1$, $k = 36$, and $(a, H) = (0.7, 0.75)$ for $(\alpha, \beta, \lambda) = (2.1, 0.8, 0.2)$ (solid), $(2.1, 1.8, 0.4)$ (dot), $(2.8, 0.4, 0.2)$ (dash), $(2.8, 0.8, 0.4)$ (dash dot).

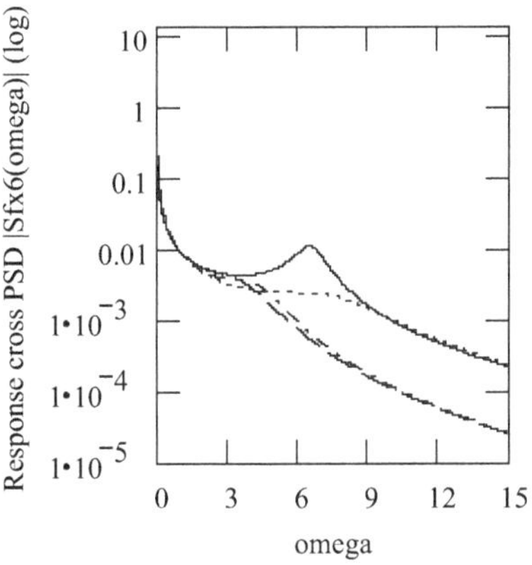


FIGURE 3.23 Plots of $|S_{fx6}(\omega)|$ in log when $m = 1$, $c = 0.1$, $k = 36$, and $(a, H) = (0.7, 0.75)$ for $(\alpha, \beta, \lambda) = (2.1, 0.8, 0.2)$ (solid), $(2.1, 1.8, 0.4)$ (dot), $(2.8, 0.4, 0.2)$ (dash), $(2.8, 0.8, 0.4)$ (dash dot).

## Theorem 3.13 (PSD response VII)

Denote by $S_{xx7}(\omega)$ the PSD of $x_7(t)$. Then,

$$S_{xx7}(\omega) = \frac{1}{k^2} \frac{-V_H H(2H-1)\sin\left[a\pi(H-1)\right]\Gamma(2aH-2a+1)|\omega|^{-2a(H-1)-1}}{\left[\omega^{\lambda}\cos\frac{\lambda\pi}{2} - \gamma\left(1-2\varsigma\omega_n\omega^{\beta-2}\cos\frac{\beta\pi}{2}\right)\right]^2 + \gamma^2\left(2\varsigma\omega^{\beta-1}\sin\frac{\beta\pi}{2} + \omega_n\omega^{\lambda-1}\sin\frac{\lambda\pi}{2}\right)^2}. \tag{3.45}$$

*Proof.* Using $S_{xx7}(\omega) = S_{ff}(\omega)|H_7(\omega)|^2$, where $H_7(\omega)$ is expressed by (Chapter 7 in Volume I or Li [2])

$$H_7(\omega) = \frac{1}{k\begin{bmatrix} \omega^\lambda \cos\frac{\lambda\pi}{2} - \gamma\left(1 - 2\varsigma\omega_n\omega^{\beta-2}\cos\frac{\beta\pi}{2}\right) \\ +i\gamma\left(2\varsigma\omega^{\beta-1}\sin\frac{\beta\pi}{2} + \omega_n\omega^{\lambda-1}\sin\frac{\lambda\pi}{2}\right)\end{bmatrix}}, \tag{3.46}$$

we have (3.45). The proof completes.

**Theorem 3.14 (cross-PSD response VII)**

Let $S_{fx7}(\omega)$ be the cross-PSD between $f(t)$ and $x_7(t)$. Then,

$$S_{fx7}(\omega) = \frac{-V_H H(2H-1)\sin\left[a\pi(H-1)\right]\Gamma(2aH-2a+1)|\omega|^{-2a(H-1)-1}}{k\begin{bmatrix} \omega^\lambda \cos\frac{\lambda\pi}{2} - \gamma\left(1 - 2\varsigma\omega_n\omega^{\beta-2}\cos\frac{\beta\pi}{2}\right) \\ +i\gamma\left(2\varsigma\omega^{\beta-1}\sin\frac{\beta\pi}{2} + \omega_n\omega^{\lambda-1}\sin\frac{\lambda\pi}{2}\right)\end{bmatrix}}. \tag{3.47}$$

*Proof.* Doing the operation of $S_{fx7}(\omega) = S_{ff}(\omega)H_7(\omega)$ with the consideration of (3.46) yields (3.47). The proof ends.

In the time domain,

$$r_{xx7}(\tau) = r_{ff}(\tau) * h_7(\tau) * h_7(-\tau), \tag{3.48}$$

where $r_{xx7}(\tau)$ is the ACF of $x_7(t)$ and $h_7(\tau)$ is the impulse response of a class VII fractional vibrator. It is in fact the ACF response. Additionally,

$$r_{fx7}(\tau) = r_{ff}(\tau) * h_7(\tau), \tag{3.49}$$

where $r_{fx7}(\tau)$ is the cross-correlation between $f(t)$ and $x_7(t)$. It is the cross-correlation response. In (3.48) and (3.49),

$$h_7(t) = e^{-\varsigma_{\mathrm{eq7}}\omega_{\mathrm{eqn7}}t}\frac{1}{m_{\mathrm{eq7}}\omega_{\mathrm{eqd7}}}\sin\omega_{\mathrm{eqd7}}t, \quad t \geq 0. \tag{3.50}$$

Refer to Chapter 7 in Volume I or Li [2] for the expressions of $\zeta_{eq7}$, $m_{eq7}$, $\omega_{eqn7}$, and $\omega_{eqd7}$ in (3.50).

### 3.8.2 Effect of ($\beta$, $\lambda$) on Responses

Figure 3.24 is used to indicate some plots of $|H_7(\omega)|^2$, Figure 3.25 is utilized for some plots of $S_{xx7}(\omega)$, and Figure 3.26 for some plots of $S_{fx7}(\omega)$. Figures 3.25 and 3.26 indicate that the effect of ($\beta$, $\lambda$) on $x_7(t)$ driven by gfGn is significant.

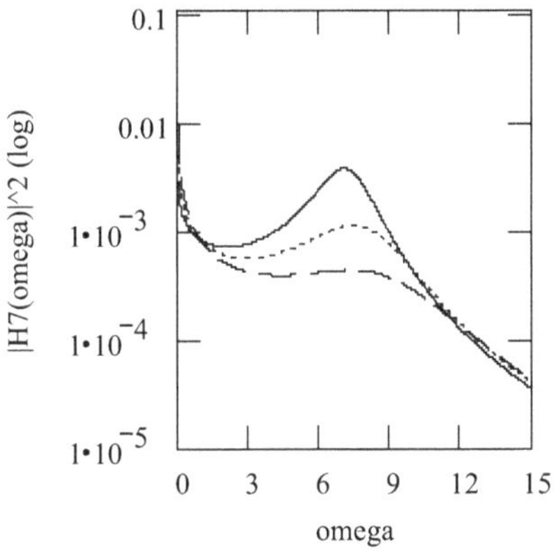


FIGURE 3.24 Plots of $|H_7(\omega)|^2$ (log) when $m = 1$, $c = 0.1$, and $k = 36$, for ($\beta$, $\lambda$) = (0.2, 0.2) (solid), (0.2, 0.3) (dot), (0.2, 0.4) (dash).

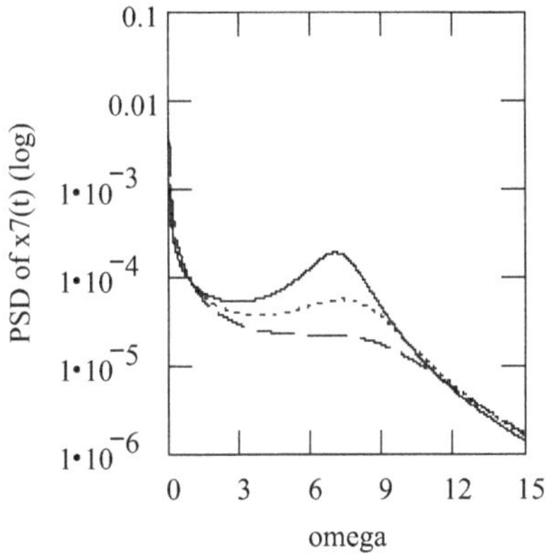


FIGURE 3.25 Plots of response PSD $S_{xx7}(\omega)$ in log when $m = 1$, $c = 0.1$, $k = 36$, and ($a$, $H$) = (0.7, 0.55) for ($\beta$, $\lambda$) = (0.2, 0.2) (solid), (0.2, 0.3) (dot), (0.2, 0.4) (dash).

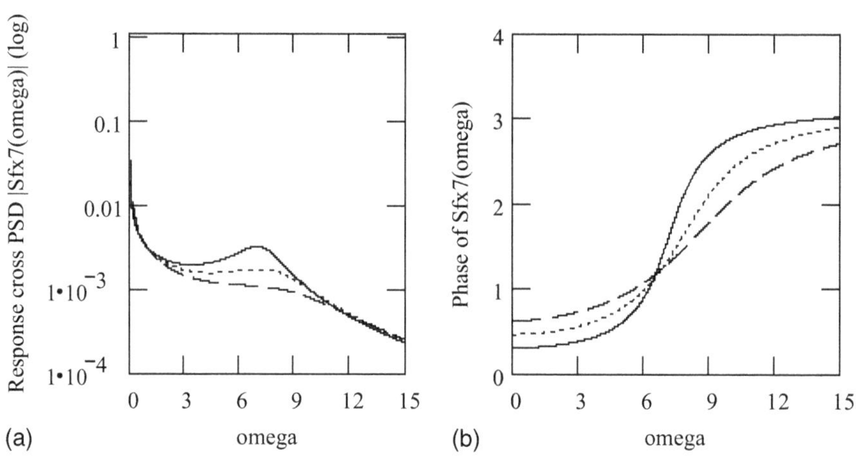


FIGURE 3.26 Plots of cross-PSD response $S_{fx7}(\omega)$ when $m = 1$, $c = 0.1$, $k = 36$, and $(a, H) = (0.7, 0.55)$ for $(\beta, \lambda) = (0.2, 0.2)$ (solid), (0.2, 0.3) (dot), (0.2, 0.4) (dash). (a). $|S_{fx7}(\omega)|$. (b). Phase of $S_{fx7}(\omega)$.

## 3.9 SUMMARY

We have presented the analytic expressions of the PSD and cross-PSD responses of seven classes of fractional vibrators under the excitation of gfGn in Theorems 3.1–3.14, respectively. The fractional orders of seven classes of vibration systems, namely, $\alpha$ or $\beta$ or $\lambda$, may considerably affect the responses. One thing worth noting is that the statistical dependences (long-range dependence or short-range dependence) follow the excitation force gfGn.

## 3.10 EXERCISES

3.1. Let

$$S_{ff}(\omega) = -V_H H(2H-1)\sin[a\pi(H-1)]\Gamma(2aH-2a+1)|\omega|^{-2a(H-1)-1},$$

where $0 < H < 1$, $0 < a \leq 1$, and $V_H = \Gamma(1-2H)\dfrac{\cos \pi H}{\pi H}$. Find the inverse Fourier transform $S_{ff}(\omega)$.

3.2. Let

$$h_2(t) = \frac{e^{-\frac{\varsigma\omega_n\omega^{\beta-1}\sin\frac{\beta\pi}{2}}{1-2\zeta\omega_n\omega^{\beta-2}\cos\frac{\beta\pi}{2}}t}\sin\dfrac{\omega_n\sqrt{1-\dfrac{\varsigma^2\omega^{2(\beta-1)}\sin^2\frac{\beta\pi}{2}}{1-2\zeta\omega_n\omega^{\beta-2}\cos\frac{\beta\pi}{2}}}}{\sqrt{1-2\zeta\omega_n\omega^{\beta-2}\cos\frac{\beta\pi}{2}}}t}{\omega_n m\sqrt{1-2\zeta\omega_n\omega^{\beta-2}\cos\frac{\beta\pi}{2}}\sqrt{1-\dfrac{\varsigma^2\omega^{2(\beta-1)}\sin^2\frac{\beta\pi}{2}}{1-2\zeta\omega_n\omega^{\beta-2}\cos\frac{\beta\pi}{2}}}}u(t).$$

Find $r_{xx2}(\tau) = r_{ff}(\tau)*h_2(\tau)*h_2(-\tau)$, where $r_{ff}(\tau) = F^{-1}[S_{ff}(\omega)]$ in Exercise 3.1.

3.3. Find $r_{fx2}(\tau) = r_{ff}(\tau)*h_2(\tau)$, where $r_{ff}(\tau) = F^{-1}[S_{ff}(\omega)]$ in Exercise 3.1.

3.4. Find the condition for $\int_{-\infty}^{\infty} r_{xx2}(\tau)d\tau = \infty$.

3.5. Find the condition for $\int_{-\infty}^{\infty} r_{xx2}(\tau)d\tau < \infty$.

3.6. Following Chapter 7 in Volume I, we denote

$$h_4(t) = e^{-\frac{m\omega^{\alpha-1}\sin\frac{\alpha\pi}{2}+k\omega^{\lambda-1}\sin\frac{\lambda\pi}{2}}{2\sqrt{mk\omega^{\alpha+\lambda-2}\left|\cos\frac{\alpha\pi}{2}\right|\cos\frac{\lambda\pi}{2}}}\sqrt{\frac{\omega^{\lambda}\cos\frac{\lambda\pi}{2}}{-\omega^{\alpha-2}\cos\frac{\alpha\pi}{2}}}\omega_n t}\frac{1}{m_{\mathrm{eq4}}\omega_{\mathrm{eqd4}}}\sin\omega_{\mathrm{eqd4}}tu(t).$$

Find $r_{xx4}(\tau) = r_{ff}(\tau)*h_4(\tau)*h_4(-\tau)$, where $r_{ff}(\tau) = F^{-1}[S_{ff}(\omega)]$ in Exercise 3.1.

3.7. Find $r_{fx4}(\tau) = r_{ff}(\tau)*h_4(\tau)$, where $r_{ff}(\tau) = F^{-1}[S_{ff}(\omega)]$ in Exercise 3.1.

3.8. Find the condition for $\int_{-\infty}^{\infty} r_{xx4}(\tau)d\tau = \infty$.

3.9. Find the condition for $\int_{-\infty}^{\infty} r_{xx4}(\tau)d\tau < \infty$.

## REFERENCES


1. M. Li, *Fractional Vibrations with Applications to Euler-Bernoulli Beams*, CRC Press, Boca Raton, 2023.
2. M. Li, Analytic theory of seven classes of fractional vibrations based on elementary functions: A tutorial review, *Symmetry*, 16(9): 2024, 1202.
3. M. Li, Three classes of fractional oscillators, *Symmetry-Basel*, 10(2): 2018, 91.
4. M. Li, *Fractal Teletraffic Modeling and Delay Bounds in Computer Communications*, CRC Press, Boca Raton, 2022.
5. M. Li, Generalized fractional Gaussian noise and its application to traffic modeling, *Physica A*, 579: 2021, 1236137, 22.
6. M. Li, Fractal time series—a tutorial review, *Mathematical Problems in Engineering*, 2010, 157264, 26.
7. M. Li, Modeling autocorrelation functions of long-range dependent teletraffic series based on optimal approximation in Hilbert space-a further study, *Applied Mathematical Modelling*, 31(3): 2007, 625–631.
8. M. Pinchas and Y. Avraham, A novel clock skew estimator and its performance for the IEEE 1588v2 (PTP) case in fractional Gaussian noise/generalized fractional Gaussian noise environment, *Frontiers in Physics*, 9: 2021, 796811.
9. M. E. Sousa-Vieira and M. Fernández-Veiga, Efficient generators of the generalized fractional Gaussian noise and Cauchy processes, *Fractal and Fractional*, 7(6): 2023, 455.
10. V. N. Gorev, A. Yu. Gusev, V. I. Korniienko, and Y. I. Shedlovska, Generalized fractional Gaussian noise prediction based on the Walsh functions, *Radio Electronics, Computer Science, Control*, 3: 2023, 48.
11. O. Sheluhin, S. Rybakov, and A. Vanyushina, Modified algorithm for detecting network attacks using the fractal dimension jump estimation method in online mode, *Proceedings of Telecommunication Universities*, 8(3): 2022, 117–126. In Russian
12. V. Starchenko, Images fractal compression algorithms, *Computer-Integrated Technologies: Education, Science, Production*, 53: 2023, 215–221. In Ukrainian.
13. M. Li, PSD and cross PSD of responses of seven classes of fractional vibrations driven by fGn, fBm, fractional OU process, and von Kármán process, *Symmetry*, 16(5): 2024, 635.

CHAPTER 4

# Responses of Fractional Vibrations Driven by Fractional Brownian Motion

THIS CHAPTER CONSIDERS THE issue of a fractional Brownian motion (fBm) passing through seven classes of fractional vibration systems. The contributions given in this chapter are in twofolds. One is to bring forward the closed-form expressions of the power spectrum density (PSD) and cross-PSD responses to seven classes of fractional vibrators driven by fBm in Theorems 4.1–4.14, respectively. The other is to demonstrate that there are considerable effects of the fractional orders of seven classes of vibration systems on the responses. The responses of seven classes of fractional vibrators driven by fBm are of long-range dependence.

## 4.1 BACKGROUND

The topic of differential equations driven by the fractional Brownian motion (fBm) receives interests of researchers, see, for example, He et al. [1, 2], Liu [3], Tuan et al. [4, 5], Sharma et al. [6], Fan et al. [7], Zhang and Yuan [8], Sun et al. [9], Shahnazi-Pour et al. [10], Araya et al. [11],

DOI: 10.1201/9781003657903-4

Gairing et al. [12], Xu et al. [13], and Heydari et al. [14, 15], simply mentioning a few.

Recently, the author introduced the closed-form analytic representations of the responses to seven classes of fractional vibration systems in [16–19]. However, reports with respect to the closed-form analytic representations of the responses to seven classes of fractional vibration systems under the excitation of fBm are seldom seen. This chapter gives the contributions in the closed-form analytic representations of the PSD and cross-PSD responses to seven classes of fractional vibration systems driven by fBm.

The rest of the chapter is organized as follows. We put forward the analytic expressions of the PSD and cross-PSD responses to seven classes of fractional vibration systems driven by fBm in Sections 4.2–4.8, respectively. The summary is given in Section 4.9.

## 4.2 RESPONSES OF CLASS I FRACTIONAL VIBRATORS DRIVEN BY FBM

### 4.2.1 Computations

Consider the following motion equation of a class I fractional vibrator in the form

$$m\frac{d^{\alpha}x_1(t)}{dt^{\alpha}}+k\frac{dx_1(t)}{dt}=f(t). \tag{4.1}$$

In (4.1), $1 < \alpha < 3$, $x_1(t)$ is the response, $f(t)$ is the excitation, $m$ and $k$ are the primary mass and stiffness, respectively.

Let $f(t)$ be a driven signal of fBm in what follows. Let $S_{ff}(t, \omega)$ be the PSD of $f(t)$. Then,

$$S_{ff}(t,\omega)=\frac{V_H}{|\omega|^{2H+1}}\left(1-2^{1-2H}\cos 2\omega t\right). \tag{4.2}$$

In (4.2), $0 < H < 1$ is the Hurst parameter and $V_H=\Gamma(1-2H)\frac{\cos\pi H}{\pi H}$.

Figure 4.1 shows a plot of $S_{ff}(t, \omega)$.

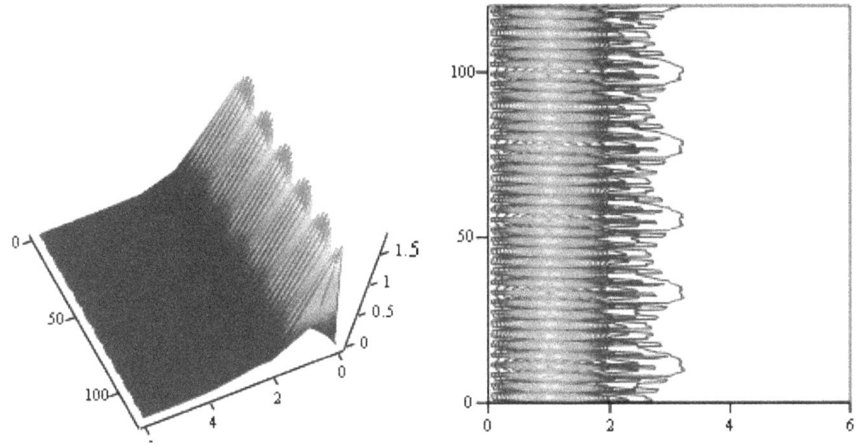


Sff
(a)

Sff
(b)

FIGURE 4.1 Plots of PSD of $S_{ff}(t, \omega)$ when writing $S_{ff}(t, \omega) =$ Sff for $\omega = 0, \ldots, 6$ and $t = 0, \ldots, 120$, and $H = 075$. (a). PSD in surface plot. (b). PSD in contour plot.

**Theorem 4.1 (PSD response I)**

Denote by $S_{xx1}(t, \omega)$ the PSD of $x_1(t)$. Then,

$$S_{xx1}(t,\omega) = \frac{\frac{V_H}{|\omega|^{2H+1}}\left(1 - 2^{1-2H}\cos 2\omega t\right)}{k^2\left[\left(1 - \frac{\omega^\alpha}{\omega_n^2}\left|\cos\frac{\alpha\pi}{2}\right|\right)^2 + \left(\frac{\omega^\alpha}{\omega_n^2}\sin\frac{\alpha\pi}{2}\right)^2\right]}, \tag{4.3}$$

where $\omega_n^2 = \dfrac{k}{m}$.

*Proof.* Note that $H_1(\omega)$ is given by (Chapter 7 in Volume I, Li [16–19])

$$H_1(\omega) = \frac{1}{k\left(1 - \frac{\omega^\alpha}{\omega_n^2}\left|\cos\frac{\alpha\pi}{2}\right| + i\frac{\omega^\alpha}{\omega_n^2}\sin\frac{\alpha\pi}{2}\right)}, \tag{4.4}$$

Thus, by considering $S_{xx1}(t, \omega) = S_{ff}(t, \omega)|H_1(\omega)|^2$, we have (4.3) (Li [20]). The proof is finished.

**Theorem 4.2 (cross-PSD response I)**

Let $S_{fx1}(t, \omega)$ be the cross PSD between $f(t)$ and $x_1(t)$. Then,

$$S_{fx1}(t,\omega)=\frac{\frac{V_H}{|\omega|^{2H+1}}\left(1-2^{1-2H}\cos 2\omega t\right)}{k\left(1-\frac{\omega^{\alpha}}{\omega_n^2}\left|\cos\frac{\alpha\pi}{2}\right|+i\frac{\omega^{\alpha}}{\omega_n^2}\sin\frac{\alpha\pi}{2}\right)}. \tag{4.5}$$

*Proof.* Doing the operation of $S_{fx1}(t, \omega) = S_{ff}(t, \omega)H_1(\omega)$ with the consideration of (4.4) yields (4.5). The proof ends.

Denote by $h_1(\tau)$ the impulse response of a class I fractional vibrator. Let $r_{xx1}(t, \tau)$ be the inverse Fourier transform of $S_{xx1}(t, \omega)$. Denote by $r_{fx1}(t, \tau)$ the inverse Fourier transform of $S_{fx1}(t, \omega)$. Then, applying the convolution theory to $S_{xx1}(t, \omega) = S_{ff}(t, \omega)|H_1(\omega)|^2$ results in the autocorrelation function (ACF) response given by

$$r_{xx1}(t, \tau) = r_{ff}(t, \tau) * h_1(\tau) * h_1(-\tau). \tag{4.6}$$

Similarly, applying the convolution theory to $S_{fx1}(t, \omega) = S_{ff}(t, \omega)H_1(\omega)$ yields the cross-correlation response in the form

$$r_{fx1}(t, \tau) = r_{ff}(t, \tau) * h_1(\tau). \tag{4.7}$$

In (4.6) and (4.7), according to Chapter 7 in Volume I or Li [16–19],

$$h_1(t)=\frac{e^{-\frac{\omega\sin\frac{\alpha\pi}{2}}{2\left|\cos\frac{\alpha\pi}{2}\right|}t}\sin\left(\frac{\omega_n}{\sqrt{\omega^{\alpha-2}\left|\cos\frac{\alpha\pi}{2}\right|}}\sqrt{1-\frac{\omega^{2\alpha}\sin^2\frac{\alpha\pi}{2}}{4\omega_n^2\left|\cos\frac{\alpha\pi}{2}\right|}}\,t\right)}{m\omega_n\sqrt{\omega^{\alpha-2}\left|\cos\frac{\alpha\pi}{2}\right|}\sqrt{1-\frac{\omega^{2\alpha}\sin^2\frac{\alpha\pi}{2}}{4\omega_n^2\left|\cos\frac{\alpha\pi}{2}\right|}}}u(t). \tag{4.8}$$

In (4.8), $u(t)$ is the unit step function.

### 4.2.2 Effect of $\alpha$ on Responses

Figure 4.2 indicates some plots of $|H_1(\omega)|^2$. Some plots of $S_{xx1}(t, \omega)$ are shown in Figure 4.3. When $\alpha = 2$, a class I fractional vibrator reduces to be a conventional damping-free vibrator. In general, there are three parameters affecting the response $S_{xx1}(t, \omega)$. From a view of system, the parameter is the fractional order $\alpha$. From a view of fractional processes, the parameter $H$ is crucial to $S_{xx1}(t, \omega)$. Figure 4.4 shows the $H$ effect on $S_{xx1}(t, \omega)$. In Figure 4.5, we illustrate a plot of the cross-PSD response $S_{fx1}(t, \omega)$. In Figure 4.6,

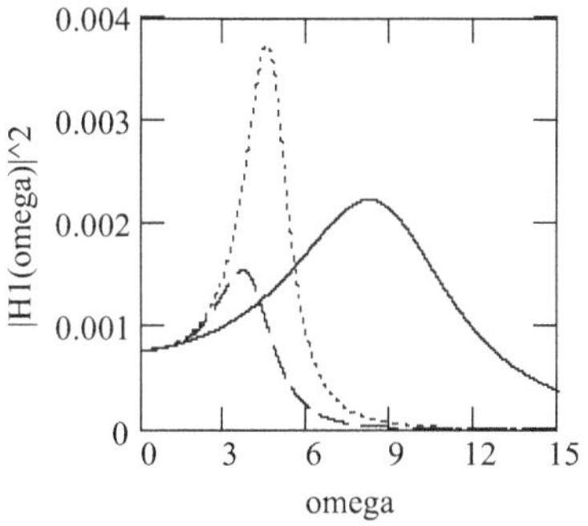


FIGURE 4.2 Plots of $|H_1(\omega)|^2$ with $\alpha = 1.6$ (solid), 2.3 (dot), 2.5 (dash) when $m = 1$ and $k = 36$.

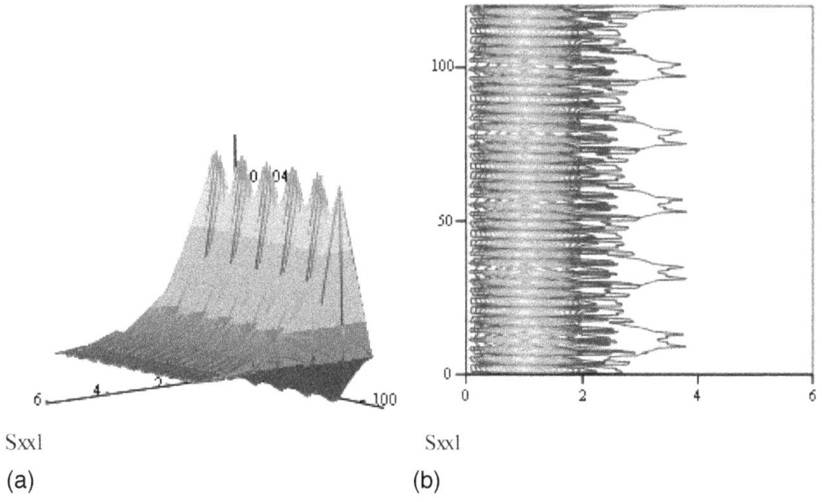


FIGURE 4.3 Plots of $S_{xx1}(t, \omega)$ for $m = 1$, $k = 36$, and $H = 0.75$ when writing $S_{xx1}(t, \omega) =$ Sxx1 for $\omega = 0, \ldots, 6$ and $t = 0, \ldots, 120$. (a). PSD in surface plot ($\alpha = 1.8$). (b). PSD in contour plot ($\alpha = 1.8$). (c). PSD in contour plot ($\alpha = 2.8$).

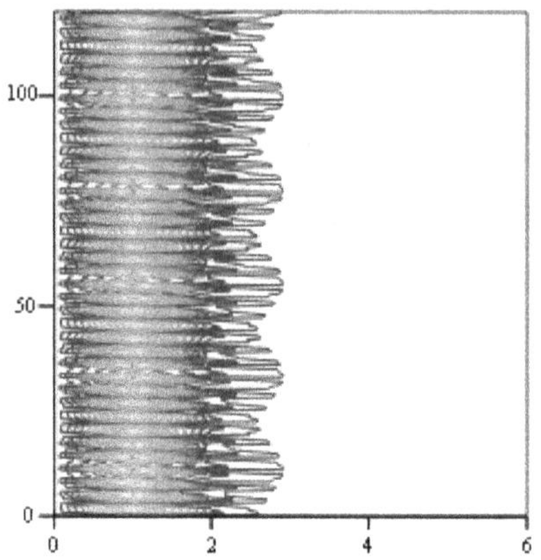


Sxx1
(c)

FIGURE 4.3 (Continued)

we show the $H$ effect on $S_{fx1}(t, \omega)$. Figures 4.3 and 4.5 indicate that there is significant effect of the fractional order $\alpha$ on the responses to class I fractional vibration systems driven by fBm.

## 4.3 RESPONSES OF CLASS II FRACTIONAL VIBRATORS DRIVEN BY FBM

### 4.3.1 Computation Methods

For a class II fractional vibrator, its motion equation is given by

$$m\frac{d^2x_2(t)}{dt^2}+c\frac{d^{\beta}x_2(t)}{dt^{\beta}}+k\frac{dx_2(t)}{dt}=f(t). \tag{4.9}$$

In (4.9), $0 < \beta < 2$, $x_2(t)$ is the response of a class II fractional vibrator and $c$ is the primary damping.

**Theorem 4.3 (PSD response II)**

Let $S_{xx2}(t, \omega)$ be the PSD of $x_2(t)$. Then,

$$S_{xx2}(t,\omega)=\frac{\frac{V_H}{|\omega|^{2H+1}}\left(1-2^{1-2H}\cos 2\omega t\right)}{k^2\left\{\left[1-\gamma^2\left(1-\frac{c}{m}\omega^{\beta-2}\cos\frac{\beta\pi}{2}\right)\right]^2+\left(\frac{2\varsigma\omega^{\beta}}{\omega_n}\sin\frac{\beta\pi}{2}\right)^2\right\}}, \tag{4.10}$$

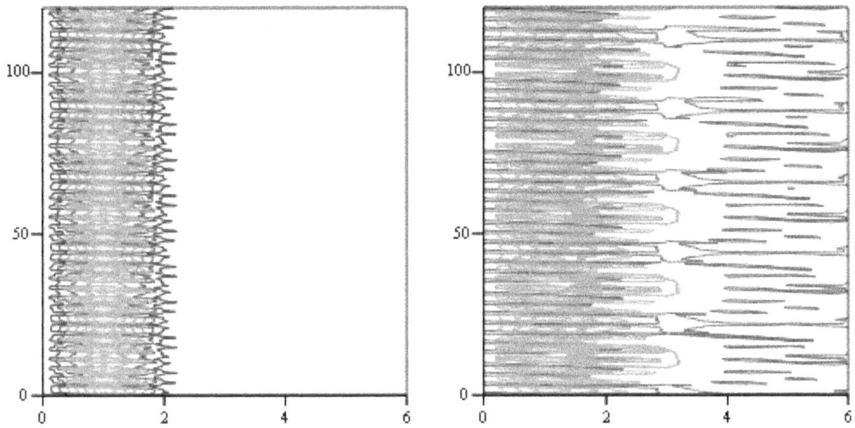


Sxxl
(a)

Sxxl
(b)

FIGURE 4.4 Observing $H$ effect on $S_{xx1}(t, \omega)$ when $\alpha = 1.8$, $m = 1$, $k = 36$ when writing $S_{xx1}(t, \omega) =$ Sxx1 for $\omega = 0, \ldots, 6$ and $t = 0, \ldots, 120$. (a). $S_{xx1}(t, \omega)$ for $H = 0.95$. (b). $S_{xx1}(t, \omega)$ for $H = 0.15$.

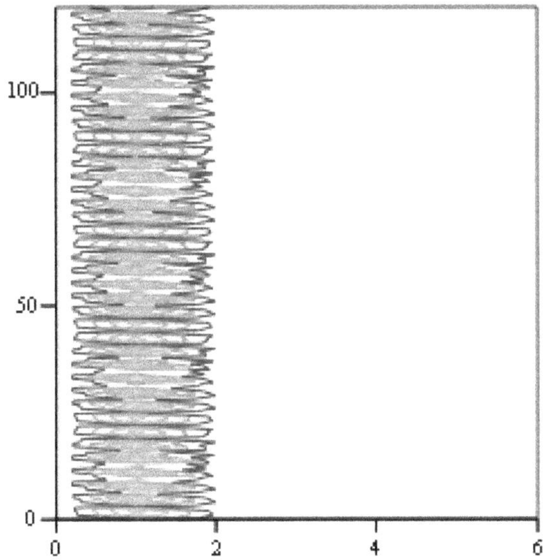


Sfx1

FIGURE 4.5 Illustration of cross-PSD response for $m = 1$, $k = 36$, and $H = 0.75$ with $\alpha = 1.8$ when writing $|S_{fx1}(t, \omega)| =$ Sfx1 for $\omega = 0, \ldots, 6$ and $t = 0, \ldots, 120$.

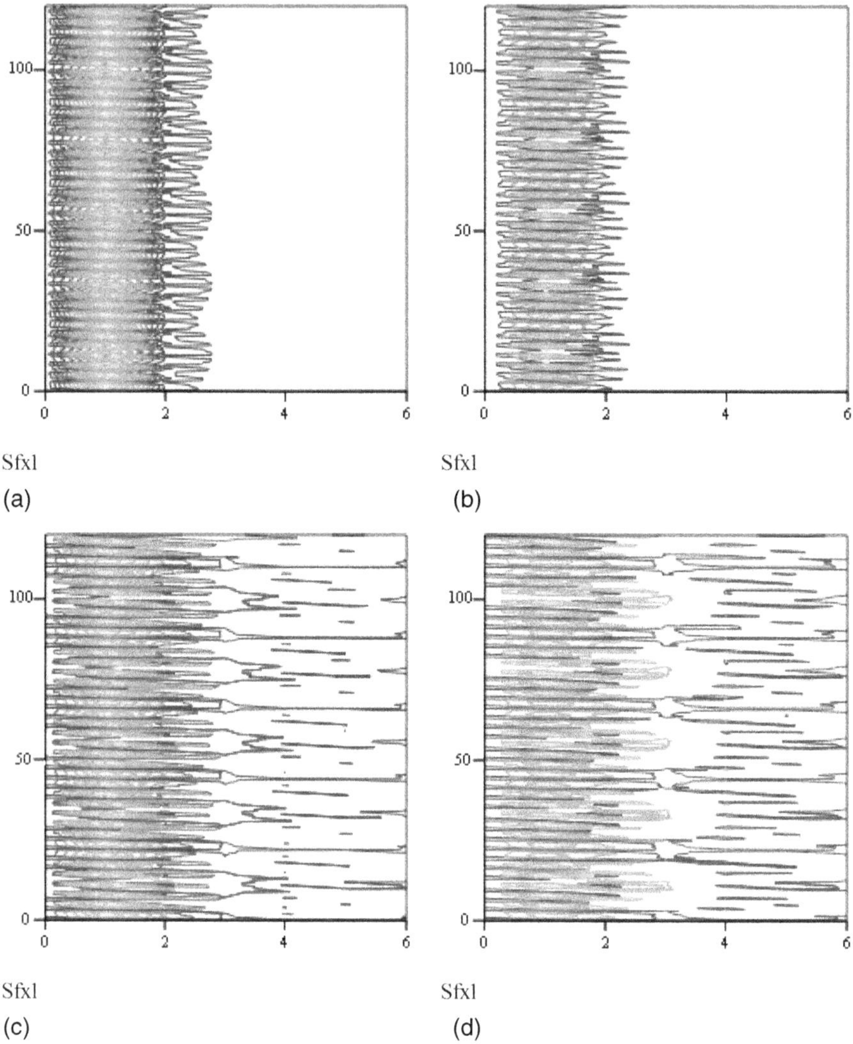


FIGURE 4.6 Observing $H$ effect on $|S_{\text{fx1}}(t, \omega)|$ when $m = 1$, $k = 36$ and $\alpha = 1.8$ when writing $|S_{\text{fx1}}(t, \omega)| = \text{Sfx1}$ for $\omega = 0, \ldots, 6$ and $t = 0, \ldots, 120$. (a). $|S_{\text{fx1}}(t, \omega)|$ for $H = 0.95$. (b). $|S_{\text{fx1}}(t, \omega)|$ for $H = 0.55$. (c). $|S_{\text{fx1}}(t, \omega)|$ for $H = 0.35$. (d). $|S_{\text{fx1}}(t, \omega)|$ for $H = 0.15$.

where $\gamma = \dfrac{\omega}{\omega_n}$.

*Proof.* Note that $S_{xx2}(t, \omega) = S_{ff}(t, \omega)|H_2(\omega)|^2$. Based on Chapter 7 in Volume I or Li [16–19], $H_2(\omega)$ is given by

$$H_2(\omega) = \frac{1/k}{1-\gamma^2\left(1-\frac{c}{m}\omega^{\beta-2}\cos\frac{\beta\pi}{2}\right)+i\frac{2\varsigma\omega^{\beta}\sin\frac{\beta\pi}{2}}{\omega_n}}. \tag{4.11}$$

Thus, we have (4.10). This finishes the proof.

**Theorem 4.4 (cross-PSD response II)**

Denote by $S_{fx2}(t, \omega)$ the cross PSD between $f(t)$ and $x_2(t)$. Then,

$$S_{fx2}(t,\omega) = \frac{\frac{V_H}{|\omega|^{2H+1}}\left(1-2^{1-2H}\cos 2\omega t\right)}{k\left[1-\gamma^2\left(1-2\varsigma\omega_n\omega^{\beta-2}\cos\frac{\beta\pi}{2}\right)+i\frac{2\varsigma\omega^{\beta}}{\omega_n}\sin\frac{\beta\pi}{2}\right]}. \tag{4.12}$$

*Proof.* Because of $S_{fx2}(t, \omega) = S_{ff}(t, \omega)H_2(\omega)$ and (4.11), (4.12) holds. The proof ends.

Let $h_2(\tau)$ be the impulse response of a class II fractional vibrator. Let $r_{xx2}(t, \tau)$ be the ACF of $x_2(t)$. Denote by $r_{fx2}(t, \tau)$ the cross-correlation between the excitation $f(t)$ and the response $x_2(t)$. According to the convolution theory with respect to $S_{xx2}(t, \omega) = S_{ff}(t, \omega)|H_2(\omega)|^2$, we have

$$r_{xx2}(t, \tau) = r_{ff}(t, \tau) * h_2(\tau) * h_2(-\tau). \tag{4.13}$$

Similarly, taking into account the convolution theory with respect to $S_{fx2}(t, \omega) = S_{ff}(t, \omega)H_2(\omega)$ produces

$$r_{fx2}(t, \tau) = r_{ff}(t, \tau) * h_2(\tau). \tag{4.14}$$

In (4.13) and (4.14), according to Chapter 7 in Volume I or Li [16–19], $h_2(t)$ is expressed by

$$h_2(t) = \frac{e^{-\frac{\varsigma\omega_n\omega^{\beta-1}\sin\frac{\beta\pi}{2}}{1-\frac{c}{m}\omega^{\beta-2}\cos\frac{\beta\pi}{2}}t}\sin\frac{\omega_n\sqrt{1-\frac{\varsigma^2\omega^{2(\beta-1)}\sin^2\frac{\beta\pi}{2}}{1-\frac{c}{m}\omega^{\beta-2}\cos\frac{\beta\pi}{2}}}}{\sqrt{1-\frac{c}{m}\omega^{\beta-2}\cos\frac{\beta\pi}{2}}}t}{\omega_n m\sqrt{1-\frac{c}{m}\omega^{\beta-2}\cos\frac{\beta\pi}{2}}\sqrt{1-\frac{\varsigma^2\omega^{2(\beta-1)}\sin^2\frac{\beta\pi}{2}}{1-\frac{c}{m}\omega^{\beta-2}\cos\frac{\beta\pi}{2}}}}u(t). \tag{4.15}$$

### 4.3.2 Effect of $\beta$ on Responses

Some plots of $|H_2(\omega)|^2$ are indicated in Figure 4.7. Figure 4.8 shows some plots of $S_{xx2}(t, \omega)$. Figure 4.9 shows the $H$ effect on $S_{xx2}(t, \omega)$. Figure 4.10 illustrates some plots of cross-PSD response $S_{fx2}(t, \omega)$. We use Figure 4.11 to exhibit the effect of $\beta$ on $|S_{fx2}(t, \omega)|$. Figures 4.8–4.11 show that the effect of $\beta$ and $H$ on the responses to class II fractional vibration systems driven by fBm is considerable.

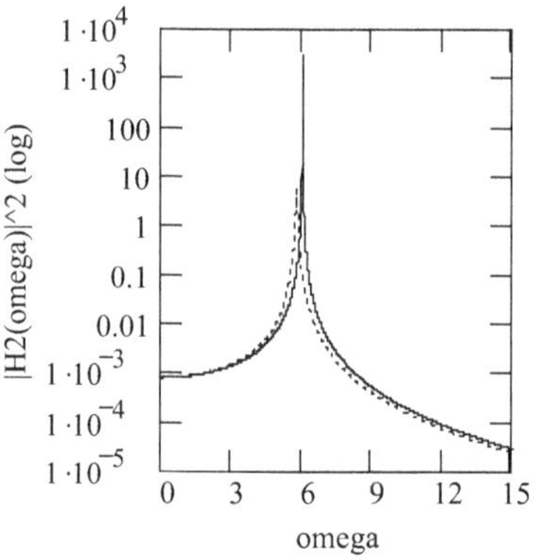


FIGURE 4.7 Plots of $|H_2(\omega)|^2$ (log) with $\beta = 0.1$ (solid), 1.9 (dot), when $m = 1$, $c = 0.1$, and $k = 36$.

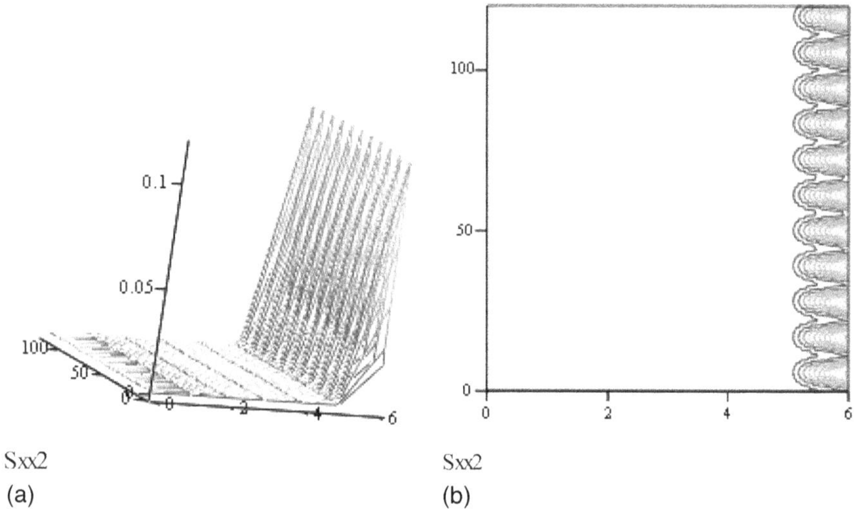


FIGURE 4.8 Plots of $S_{xx2}(t, \omega)$ with $m = 1$, $c = 0.1$, and $k = 36$ for $H = 0.75$ when writing $S_{xx2}(t, \omega) =$ Sxx2 for $\omega = 0, \ldots, 6$ and $t = 0, \ldots, 120$. (a). Surface plot of $S_{xx2}(t, \omega)$ for $\beta = 0.8$. (b). Contour plot of $S_{xx2}(t, \omega)$ for $\beta = 0.8$. (c). Surface plot of $S_{xx2}(t, \omega)$ for $\beta = 1.8$. (b). Contour plot of $S_{xx2}(t, \omega)$ for $\beta = 1.8$.

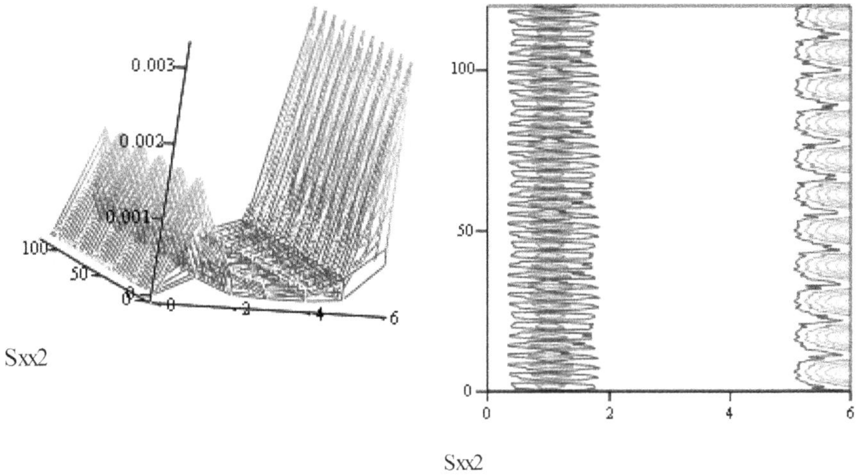


(c) (d)

FIGURE 4.8 (Continued)

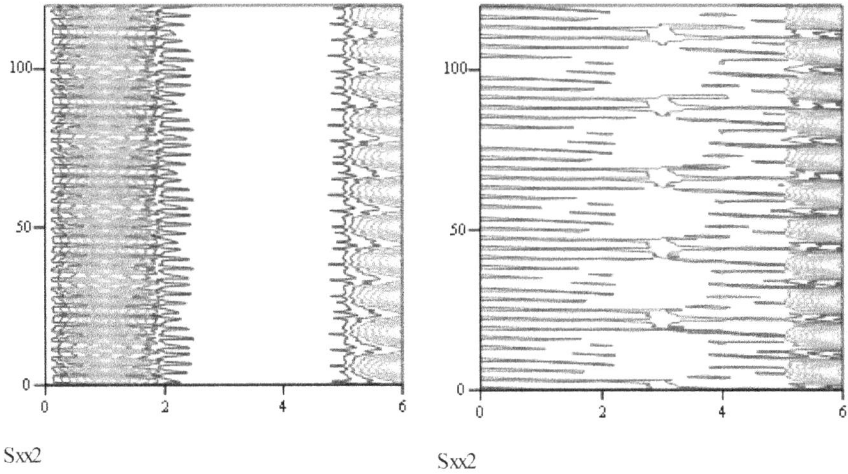


(a) (b)

FIGURE 4.9 Plots of $H$ effect on $S_{xx2}(t, \omega)$ with $\beta = 1.8$, $m = 1$, $c = 0.1$, and $k = 36$ when writing $S_{xx2}(t, \omega) = \text{Sxx2}$ for $\omega = 0, \ldots, 6$ and $t = 0, \ldots, 120$. (a). $S_{xx2}(t, \omega)$ for $H = 0.95$. (b). $S_{xx2}(t, \omega)$ for $H = 0.15$.

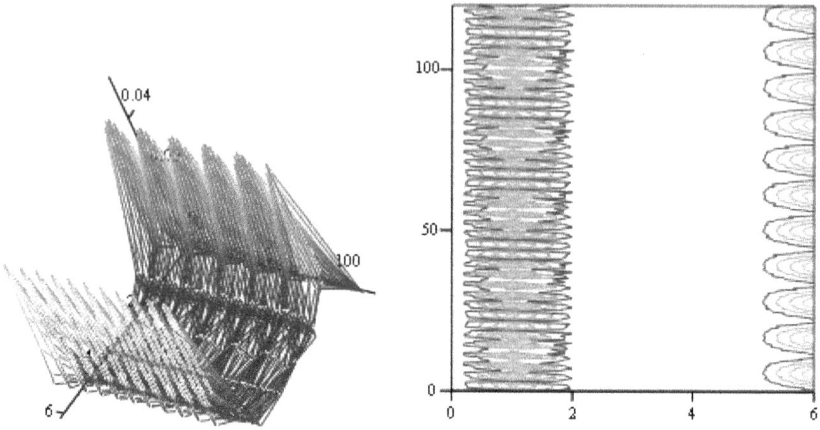


Sfx2
(a)

Sfx2
(b)

FIGURE 4.10 Plots of $|S_{fx2}(t, \omega)|$ when $m = 1$, $c = 0.1$, $k = 36$, and $(\beta, H) = (0.8, 0.75)$ when writing $|S_{fx2}(t, \omega)| =$ Sfx2 for $\omega = 0, \ldots, 6$ and $t = 0, \ldots, 120$. (a). $|S_{fx2}(t, \omega)|$ in surface plot. (b). $S_{fx2}(t, \omega)$ in contour plot.

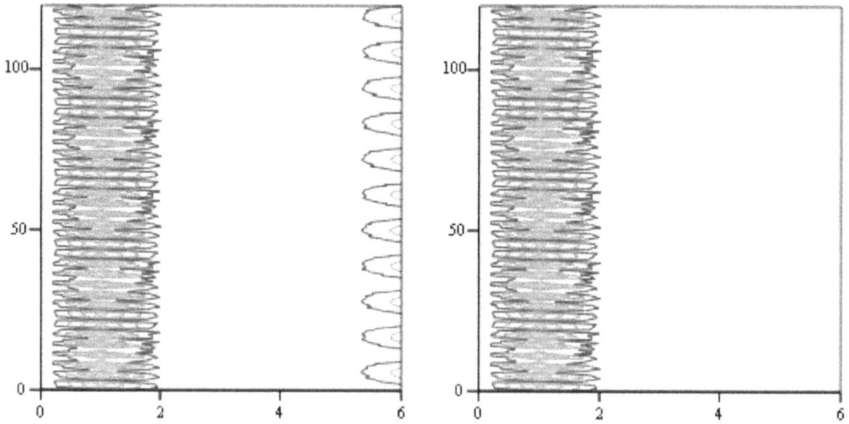


Sfx2
(a)

Sfx2
(b)

FIGURE 4.11 Response $|S_{fx2}(t, \omega)|$ when $m = 1$, $c = 0.1$, $k = 36$, and $H = 0.75$ when writing $|S_{fx2}(t, \omega)| =$ Sfx2 for $\omega = 0, \ldots, 6$ and $t = 0, \ldots, 120$. (a). $|S_{fx2}(t, \omega)|$ for $\beta = 1.2$. (b). $|S_{fx2}(t, \omega)|$ for $\beta = 1.8$.

## 4.4 RESPONSES OF CLASS III FRACTIONAL VIBRATORS DRIVEN BY FBM

### 4.4.1 Computations

The motion equation of a class III fractional vibrator is given by

$$m\frac{d^{\alpha}x_3(t)}{dt^{\alpha}}+c\frac{d^{\beta}x_3(t)}{dt^{\beta}}+kx_3(t)=f(t). \tag{4.16}$$

In (4.16), $x_3(t)$ is the response of a class III fractional vibrator.

**Theorem 4.5 (PSD response III)**

Let $S_{xx3}(t, \omega)$ be the PSD of $x_3(t)$. Then,

$$S_{xx3}(t,\omega)=\frac{\frac{V_H}{|\omega|^{2H+1}}\left(1-2^{1-2H}\cos 2\omega t\right)}{k^2\left\{\left[1-\gamma^2\left(\omega^{\alpha-2}\left|\cos\frac{\alpha\pi}{2}\right|-2\varsigma\omega_n\omega^{\beta-2}\cos\frac{\beta\pi}{2}\right)\right]^2+\left[\frac{\gamma\left(\omega^{\alpha-1}\sin\frac{\alpha\pi}{2}+2\varsigma\omega_n\omega^{\beta-1}\sin\frac{\beta\pi}{2}\right)}{\omega_n\left(\omega^{\alpha-2}\left|\cos\frac{\alpha\pi}{2}\right|-2\varsigma\omega_n\omega^{\beta-2}\cos\frac{\beta\pi}{2}\right)}\right]^2\right\}}. \tag{4.17}$$

*Proof.* With $S_{xx3}(t, \omega) = S_{ff}(t, \omega)|H_3(\omega)|^2$ and $H_3(\omega)$ expressed by (Chapter 7 in Volume I or Li [16–19])

$$H_3(\omega)=\frac{1/k}{1-\gamma^2\left(\omega^{\alpha-2}\left|\cos\frac{\alpha\pi}{2}\right|-2\varsigma\omega_n\omega^{\beta-2}\cos\frac{\beta\pi}{2}\right)+i\frac{\gamma\left(\omega^{\alpha-1}\sin\frac{\alpha\pi}{2}+2\varsigma\omega_n\omega^{\beta-1}\sin\frac{\beta\pi}{2}\right)}{\omega_n\left(\omega^{\alpha-2}\left|\cos\frac{\alpha\pi}{2}\right|-2\varsigma\omega_n\omega^{\beta-2}\cos\frac{\beta\pi}{2}\right)}}, \tag{4.18}$$

we have (4.17). The proof is finished.

**Theorem 4.6 (cross-PSD response III)**

Let $S_{fx3}(t, \omega)$ be the cross-PSD between $f(t)$ and $x_3(t)$. Then,

$$S_{fx3}(t,\omega)=\frac{\frac{V_H}{|\omega|^{2H+1}}\left(1-2^{1-2H}\cos 2\omega t\right)}{1-\gamma^2\left(\omega^{\alpha-2}\left|\cos\frac{\alpha\pi}{2}\right|-2\varsigma\omega_n\omega^{\beta-2}\cos\frac{\beta\pi}{2}\right)+i\frac{\gamma\left(\omega^{\alpha-1}\sin\frac{\alpha\pi}{2}+2\varsigma\omega_n\omega^{\beta-1}\sin\frac{\beta\pi}{2}\right)}{\omega_n\left(\omega^{\alpha-2}\left|\cos\frac{\alpha\pi}{2}\right|-2\varsigma\omega_n\omega^{\beta-2}\cos\frac{\beta\pi}{2}\right)}}. \tag{4.19}$$

*Proof.* Doing the operation of $S_{fx3}(t, \omega) = S_{ff}(t, \omega)H_3(\omega)$ and (4.18) yields (4.19). The proof ends.

Denote by $r_{xx3}(t, \tau)$ the ACF of $x_3(t)$. Let $r_{fx3}(t, t)$ be the cross-correlation between $f(t)$ and $x_3(t)$. According to the Wiener-Khinchin relation and the Wiener-Lee relation, we have the ACF response $r_{xx3}(t, \tau)$ expressed by

$$r_{xx3}(t, \tau) = r_{ff}(t, \tau) * h_3(\tau) * h_3(-\tau), \tag{4.20}$$

and the cross-correlation response given by

$$r_{fx3}(t, \tau) = r_{ff}(t, \tau) * h_3(\tau). \tag{4.21}$$

In (4.20) and (4.22), $h_3(t)$ is the impulse response function of a class III fractional vibrator. According to Chapter 7 in Volume I or Li [16–19],

$$h_3(t)=\frac{e^{-\frac{m\omega^{\alpha-1}\sin\frac{\alpha\pi}{2}+c\omega^{\beta-1}\sin\frac{\beta\pi}{2}}{2\sqrt{-\left(m\omega^{\alpha-2}\cos\frac{\alpha\pi}{2}+c\omega^{\beta-2}\cos\frac{\beta\pi}{2}\right)k}}\omega_{\text{eqn3}}t}\sin\omega_{\text{eqd3}}t}{-\left(m\omega^{\alpha-2}\cos\frac{\alpha\pi}{2}+c\omega^{\beta-2}\cos\frac{\beta\pi}{2}\right)\omega_{\text{eqd3}}}u(t). \tag{4.22}$$

Refer to Chapter 7 in Volume I or Li [16–19] for the expressions of $\omega_{\text{eqn3}}$ and $\omega_{\text{eqd3}}$ in (4.22).

### 4.4.2 Effect of $(\alpha, \beta)$ on Responses

Some plots of $|H_3(\omega)|^2$ are shown in Figure 4.12. Figure 4.13 indicates some plots of $S_{xx3}(\omega)$. Some plots of $|S_{fx3}(\omega)|$ are indicated in Figure 4.14. Figures 4.13 and 4.14 exhibit that the effect of $(\alpha, \beta)$ and $H$ on the responses to class III fractional vibrators driven by fBm is significant.

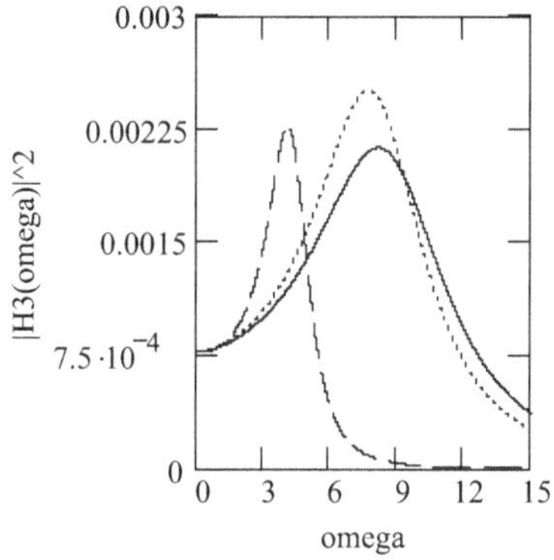


FIGURE 4.12 Plots of $|H_3(\omega)|^2$ when $m = 1$, $c = 0.1$, and $k = 36$, for $(\alpha, \beta) = (1.6, 0.8)$ (solid), $(1.6, 1.8)$ (dot), $(2.4, 0.8)$ (dash).

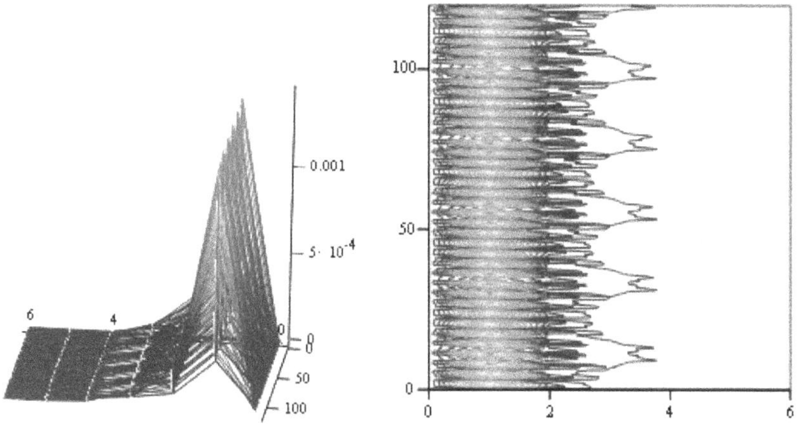


Sxx3 (a)

Sxx3 (b)

FIGURE 4.13 Plots of $S_{xx3}(t, \omega)$ with $m = 1$, $c = 0.1$, and $k = 36$ when writing $S_{xx3}(t, \omega) =$ Sxx3 for $\omega = 0, \ldots, 6$ and $t = 0, \ldots, 120$. (a). PSD in surface plot for $(\alpha, \beta) = (1.8, 0.8)$ and $H = 0.75$. (b). PSD in contour plot for $(\alpha, \beta) = (1.8, 0.8)$ and $H = 0.75$. (c). PSD in contour plot for $(\alpha, \beta) = (2.8, 1.8)$ and $H = 0.75$. (d). PSD in contour plot for $(\alpha, \beta) = (2.8, 0.8)$ and $H = 0.35$.

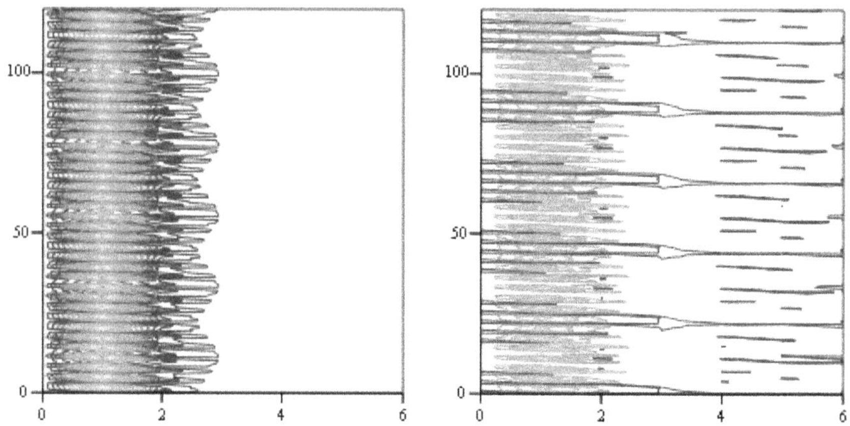


Sxx3
(c)

Sxx3
(d)

FIGURE 4.13 (Continued)

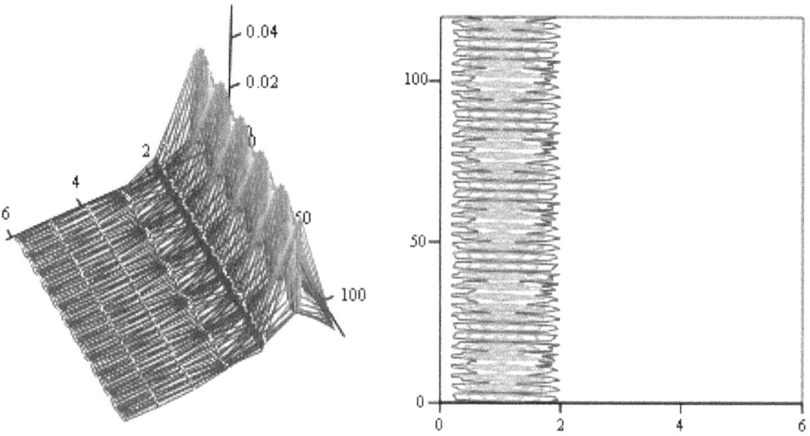


Sfx3
(a)

Sfx3
(b)

FIGURE 4.14 Plots of $|S_{fx3}(t, \omega)|$ with $m = 1$, $c = 0.1$, and $k = 36$ when writing $|S_{fx3}(t, \omega)| = \text{Sfx3}$ for $\omega = 0, \ldots, 6$ and $t = 0, \ldots, 120$. (a). Surface plot when $H = 0.75$ for $(\alpha, \beta) = (1.8, 0.8)$. (b). Contour plot when $H = 0.75$ for $(\alpha, \beta) = (1.8, 0.8)$. (c). Surface plot when $H = 0.35$ for $(\alpha, \beta) = (1.8, 0.8)$. (d). Contour plot when $H = 0.35$ for $(\alpha, \beta) = (1.8, 0.8)$.

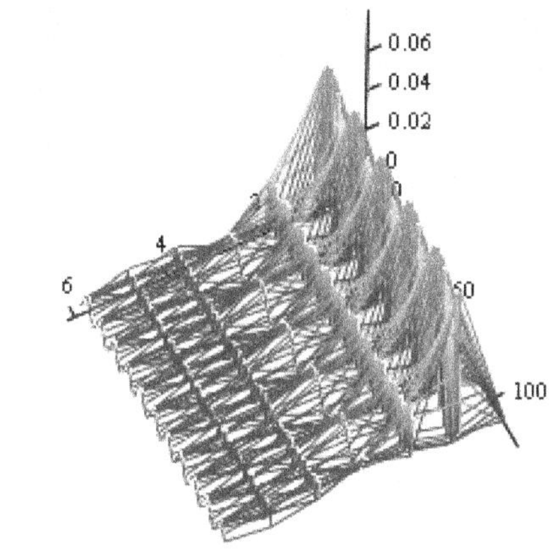


Sfx3

(c)

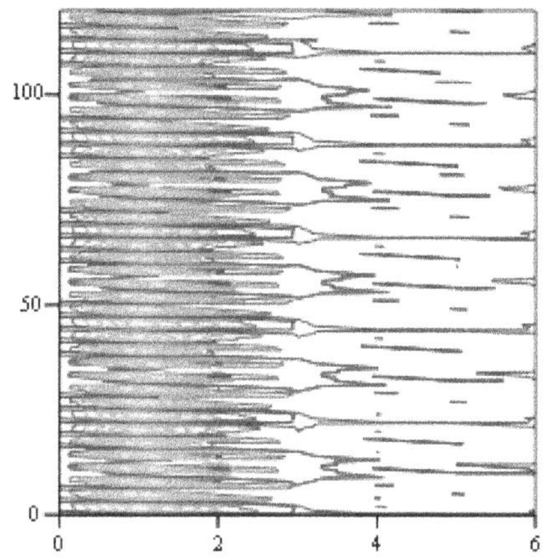


Sfx3

(d)

FIGURE 4.14 (Continued)

## 4.5 RESPONSES OF CLASS IV FRACTIONAL VIBRATORS DRIVEN BY FBM

### 4.5.1 Computations

The motion equation of a class IV fractional vibrator is given by

$$m\frac{d^{\alpha}x_4(t)}{dt^{\alpha}}+k\frac{d^{\lambda}x_4(t)}{dt^{\lambda}}=f(t). \tag{4.23}$$

In (4.23), $0 \le \lambda < 1$ and $x_4(t)$ is the response of a class IV fractional vibrator.

**Theorem 4.7 (PSD response IV)**

Let $S_{xx4}(t, \omega)$ be the PSD of $x_4(t)$. Then,

$$S_{xx4}(t,\omega)=\frac{\frac{V_H}{|\omega|^{2H+1}}\left(1-2^{1-2H}\cos 2\omega t\right)}{k^2\omega^{2\lambda}\cos^2\frac{\lambda\pi}{2}\left[\left(1-\gamma^2\frac{-\omega^{\alpha-2}\cos\frac{\alpha\pi}{2}}{\omega^{\lambda}\cos\frac{\lambda\pi}{2}}\right)^2+4\left(\gamma\frac{m\omega^{\alpha-1}\sin\frac{\alpha\pi}{2}+k\omega^{\lambda-1}\sin\frac{\lambda\pi}{2}}{2\sqrt{mk\omega^{\alpha+\lambda-2}\left|\cos\frac{\alpha\pi}{2}\right|\cos\frac{\lambda\pi}{2}}}\sqrt{\frac{-\omega^{\alpha-2}\cos\frac{\alpha\pi}{2}}{\omega^{\lambda}\cos\frac{\lambda\pi}{2}}}\right)^2\right]}. \tag{4.24}$$

*Proof.* Performing the operation of $S_{xx4}(t,\omega) = S_{ff}(t,\omega)|H_4(\omega)|^2$ and using $H_4(\omega)$ expressed by (Chapter 7 in Volume I or Li [16, 17] or Li [19])

$$H_4(\omega)=\frac{1}{k\omega^{\lambda}\cos\frac{\lambda\pi}{2}\left(1-\gamma^2\frac{-\omega^{\alpha-2}\cos\frac{\alpha\pi}{2}}{\omega^{\lambda}\cos\frac{\lambda\pi}{2}}+i2\gamma\frac{m\omega^{\alpha-1}\sin\frac{\alpha\pi}{2}+k\omega^{\lambda-1}\sin\frac{\lambda\pi}{2}}{2\sqrt{mk\omega^{\alpha+\lambda-2}\left|\cos\frac{\alpha\pi}{2}\right|\left|\cos\frac{\lambda\pi}{2}\right|}}\sqrt{\frac{-\omega^{\alpha-2}\cos\frac{\alpha\pi}{2}}{\omega^{\lambda}\cos\frac{\lambda\pi}{2}}}\right)}, \tag{4.25}$$

we have (4.24). This finishes the proof.

**Theorem 4.8 (cross-PSD response IV)**

Denote by $S_{fx4}(t,\omega)$ the cross PSD between $f(t)$ and $x_4(t)$. Then,

$$S_{fx4}(t,\omega)=\frac{\frac{V_H}{|\omega|^{2H+1}}\left(1-2^{1-2H}\cos 2\omega t\right)}{k\omega^{\lambda}\cos\frac{\lambda\pi}{2}\left(1-\gamma^2\frac{-\omega^{\alpha-2}\cos\frac{\alpha\pi}{2}}{\omega^{\lambda}\cos\frac{\lambda\pi}{2}}+i2\gamma\frac{m\omega^{\alpha-1}\sin\frac{\alpha\pi}{2}+k\omega^{\lambda-1}\sin\frac{\lambda\pi}{2}}{2\sqrt{mk\omega^{\alpha+\lambda-2}\left|\cos\frac{\alpha\pi}{2}\right|\left|\cos\frac{\lambda\pi}{2}\right|}}\sqrt{\frac{-\omega^{\alpha-2}\cos\frac{\alpha\pi}{2}}{\omega^{\lambda}\cos\frac{\lambda\pi}{2}}}\right)}. \tag{4.26}$$

*Proof.* Doing the operation of $S_{fx4}(t,\omega) = S_{ff}(t,\omega)H_4(\omega)$ with the expression of $H_4(\omega)$ in (4.25) produces (4.26). This finishes the proof.

Doing the inverse Fourier transform on both sides of $S_{xx4}(t,\omega) = S_{ff}(t,\omega)|H_4(\omega)|^2$ yields the ACF response

$$r_{xx4}(t,\tau) = r_{ff}(t,\tau)*h_4(\tau)*h_4(-\tau), \tag{4.27}$$

where $h_4(\tau)$ is the impulse response of a class IV fractional vibrator. Performing the inverse Fourier transform on both sides of $S_{fx4}(t,\omega) = S_{ff}(t,\omega)H_4(\omega)$ results in the cross-correlation response

$$r_{fx4}(t,\tau) = r_{ff}(t,\tau)*h_4(\tau). \tag{4.28}$$

In (4.27) and (4.28), $h_4(t)$ is given by (Chapter 7 in Volume I or Li [16, 17] or Li [19])

$$h_4(t) = e^{-\frac{m\omega^{\alpha-1}\sin\frac{\alpha\pi}{2}+k\omega^{\lambda-1}\sin\frac{\lambda\pi}{2}}{2\sqrt{mk\omega^{\alpha+\lambda-2}\left|\cos\frac{\alpha\pi}{2}\right|\cos\frac{\lambda\pi}{2}}}\sqrt{\frac{\omega^{\lambda}\cos\frac{\lambda\pi}{2}}{-\omega^{\alpha-2}\cos\frac{\alpha\pi}{2}}}\omega_n t} \frac{1}{m_{\mathrm{eq4}}\omega_{\mathrm{eqd4}}}\sin\omega_{\mathrm{eqd4}}tu(t). \tag{4.29}$$

Refer to Chapter 7 in Volume I or Li [16, 17] or Li [19] for the expressions of $m_{\mathrm{eq4}}$ and $\omega_{\mathrm{eqd4}}$ in (4.29).

### 4.5.2 Effect of $(\alpha, \lambda)$ on Responses

Figure 4.15 illustrates some plots of $|H_4(\omega)|^2$. Figure 4.16 shows some plots of $S_{xx4}(t, \omega)$. Some plots of $|S_{fx4}(t, \omega)|$ are in Figure 4.17. Figures 4.16 and 4.17 show that there is noticeable effect of $(\alpha, \lambda)$ on the responses to class IV fractional vibrators driven by fBm.

## 4.6 RESPONSES OF CLASS V FRACTIONAL VIBRATORS DRIVEN BY FBM

### 4.6.1 Computation Methods

Consider the motion equation of a class V fractional vibrator:

$$m\frac{d^2x_5(t)}{dt^2} + k\frac{d^{\lambda}x_5(t)}{dt^{\lambda}} = f(t). \tag{4.30}$$

In (4.30), $x_5(t)$ is the response of a class V fractional vibrator.

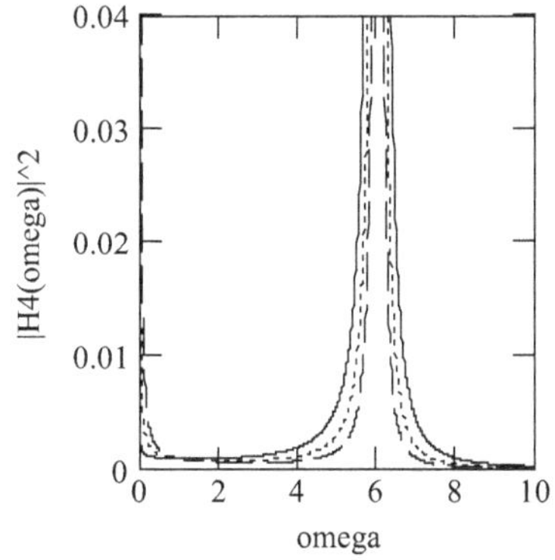


FIGURE 4.15 Plots of $|H_4(\omega)|^2$ when $m = 1$, $c = 0$, and $k = 36$, for $(\alpha, \lambda) = (2.1, 0.1)$ (solid), (2.3, 0.3) (dot), (2.5, 0.5) (dash).

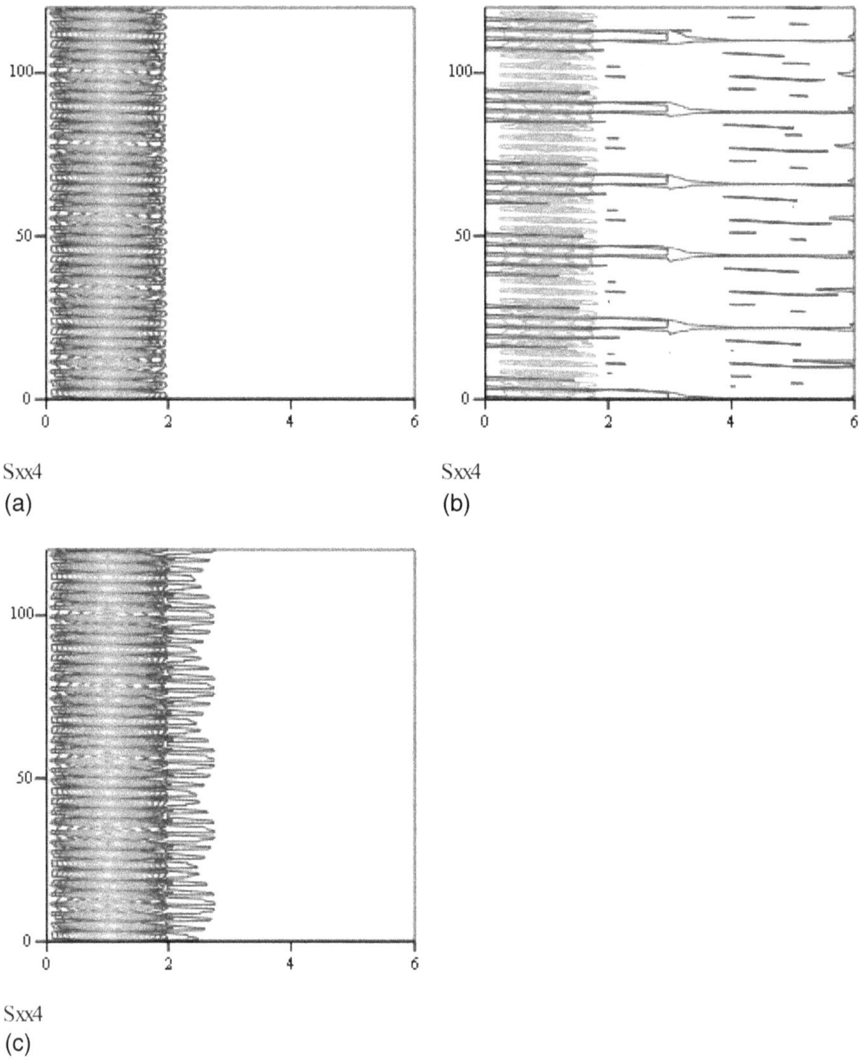


FIGURE 4.16 Plots of $S_{xx4}(t, \omega)$ with $m = 1$, $c = 0$, and $k = 36$ when writing $S_{xx4}(t, \omega) = \text{Sxx4}$ for $\omega = 0, \ldots, 6$ and $t = 0, \ldots, 120$. (a). $S_{xx4}(t, \omega)$ when $H = 0.75$ for $(\alpha, \lambda) = (1.8, 0.8)$. (b). $S_{xx4}(t, \omega)$ when $H = 0.35$ for $(\alpha, \lambda) = (1.8, 0.8)$. (c). $S_{xx4}(t, \omega)$ when $H = 0.75$ for $(\alpha, \lambda) = (1.8, 0.2)$.

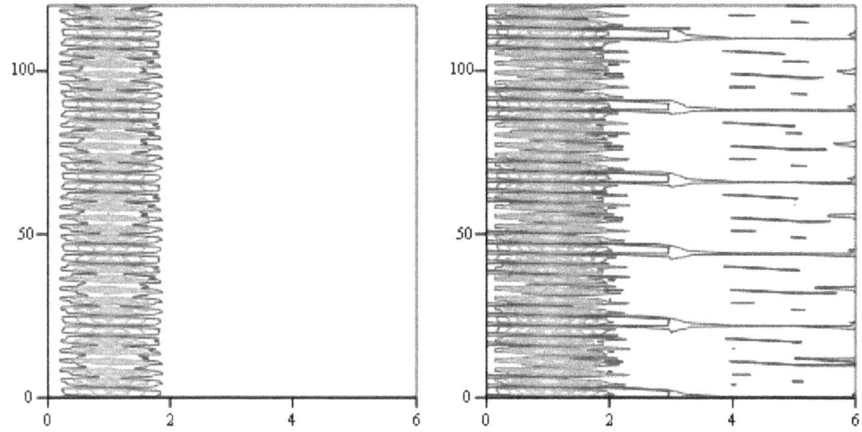


Sfx4
(a)

Sfx4
(b)

FIGURE 4.17 Plots of $|S_{fx4}(t, \omega)|$ with $m = 1$, $c = 0$, and $k = 36$ when writing $|S_{fx4}(t, \omega)| = \text{Sfx4}$ for $\omega = 0, \ldots, 6$ and $t = 0, \ldots, 120$. (a). $|S_{fx4}(t, \omega)|$ for $(\alpha, \lambda) = (1.8, 0.8)$ and $H = 0.75$. (b). $|S_{fx4}(t, \omega)|$ for $(\alpha, \lambda) = (1.8, 0.8)$ and $H = 0.95$.

**Theorem 4.9 (PSD response V)**

Denote by $S_{xx5}(t, \omega)$ the PSD of $x_5(t)$. Then,

$$S_{xx5}(t,\omega) = \frac{\dfrac{V_H}{|\omega|^{2H+1}}\left(1-2^{1-2H}\cos 2\omega t\right)}{k^2\omega^{2\lambda}\cos^2\dfrac{\lambda\pi}{2}\left[\left(1-\dfrac{\gamma^2}{\omega^\lambda\cos\dfrac{\lambda\pi}{2}}\right)^2 + 4\gamma^2\left(\dfrac{k\omega^{\lambda-1}\sin\dfrac{\lambda\pi}{2}}{2\sqrt{mk\omega^\lambda\cos\dfrac{\lambda\pi}{2}}}\sqrt{\dfrac{1}{\omega^\lambda\cos\dfrac{\lambda\pi}{2}}}\right)^2\right]}. \tag{4.31}$$

*Proof.* Doing $S_{xx5}(t, \omega) = S_{ff}(t, \omega)|H_5(\omega)|^2$, where (Chapter 7 in Volume I or Li [16, 17] or Li [19])

$$H_5(\omega)=\frac{1}{k\omega^\lambda\cos\frac{\lambda\pi}{2}\left(1-\frac{\gamma^2}{\frac{\omega^\lambda\cos\frac{\lambda\pi}{2}}{\sqrt{\frac{1}{\omega^\lambda\cos\frac{\lambda\pi}{2}}}}}+i2\gamma\frac{k\omega^{\lambda-1}\sin\frac{\lambda\pi}{2}}{2\sqrt{mk\omega^\lambda\cos\frac{\lambda\pi}{2}}}\right)}, \tag{4.32}$$

yields (4.31). The proof ends.

**Theorem 4.10 (cross-PSD response V)**

Let $S_{fx5}(t, \omega)$ be the cross PSD between $f(t)$ and $x_5(t)$. Then,

$$S_{fx5}(t,\omega)=\frac{\frac{V_H}{|\omega|^{2H+1}}\left(1-2^{1-2H}\cos 2\omega t\right)}{k\omega^\lambda\cos\frac{\lambda\pi}{2}\left(1-\frac{\gamma^2}{\frac{\omega^\lambda\cos\frac{\lambda\pi}{2}}{\sqrt{\frac{1}{\omega^\lambda\cos\frac{\lambda\pi}{2}}}}}+i2\gamma\frac{k\omega^{\lambda-1}\sin\frac{\lambda\pi}{2}}{2\sqrt{mk\omega^\lambda\cos\frac{\lambda\pi}{2}}}\right)}. \tag{4.33}$$

*Proof.* With $S_{fx5}(t, \omega) = S_{ff}(t, \omega)H_5(\omega)$ and (4.32), we have (4.33). This finishes the proof.

Doing the inverse Fourier transform on both sides of $S_{xx5}(t, \omega) = S_{ff}(t, \omega)$ $|H_5(\omega)|^2$ produces the ACF response in the form

$$r_{xx5}(t, \tau) = r_{ff}(t, \tau)*h_5(\tau)*h_5(-\tau), \tag{4.34}$$

where $h_5(\tau)$ is the impulse response of a class V fractional vibrator. Considering the convolution theory with respect to $S_{fx5}(t, \omega) = S_{ff}(t, \omega)$ $H_5(\omega)$ yields the cross-correlation response given by

$$r_{fx5}(t, \tau) = r_{ff}(t, \tau)*h_5(\tau), \tag{4.35}$$

where $r_{fx5}(t, \tau)$ is the cross-correlation between $f(t)$ and $x_5(t)$. In (4.34) and (4.35), following Chapter 7 in Volume I or Li [16, 17] or Li [19],

$$h_5(t) = e^{-\frac{k\omega^{\lambda-1}\sin\frac{\lambda\pi}{2}}{2\sqrt{mk\omega^{\lambda}\cos\frac{\lambda\pi}{2}}}\sqrt{\omega^{\lambda}\cos\frac{\lambda\pi}{2}}\omega_n t} \frac{1}{m\omega_{\text{eqd}5}} \sin\omega_{\text{eqd}5} t u(t). \tag{4.36}$$

Refer to Chapter 7 in Volume I or Li [16, 17] or Li [19] for the expression of $\omega_{\text{eqd}5}$ in (4.36).

### 4.6.2 Effect of $\lambda$ on Responses

Figure 4.18 illustrates some plots of $|H_5(\omega)|^2$, Figure 4.19 is for some plots of $S_{xx5}(t, \omega)$. Figure 4.20 is for some plots of $|S_{fx5}(t, \omega)|$. Figures 4.19 and 4.20 show that the effect of $\lambda$ on the responses to class V fractional vibration systems driven by fBm is considerable.

## 4.7 RESPONSES OF CLASS VI FRACTIONAL VIBRATORS DRIVEN BY FBM

### 4.7.1 Computations

Consider the motion equation of a class VI fractional vibrator in the form

$$m\frac{d^{\alpha}x_6(t)}{dt^{\alpha}} + c\frac{d^{\beta}x_6(t)}{dt^{\beta}} + k\frac{d^{\lambda}x_6(t)}{dt^{\lambda}} = f(t). \tag{4.37}$$

In (4.37), $x_6(t)$ is the response of a class VI fractional vibrator.

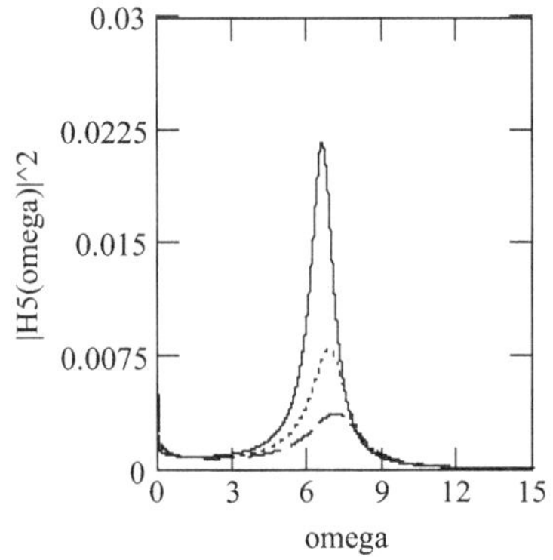


FIGURE 4.18 Plots of $|H_5(\omega)|^2$ when $m = 1$, $c = 0$, and $k = 36$, for $\lambda = 0.2$ (solid), 0.10 (dot), 0.15 (dash), 0.20 (dash dot).

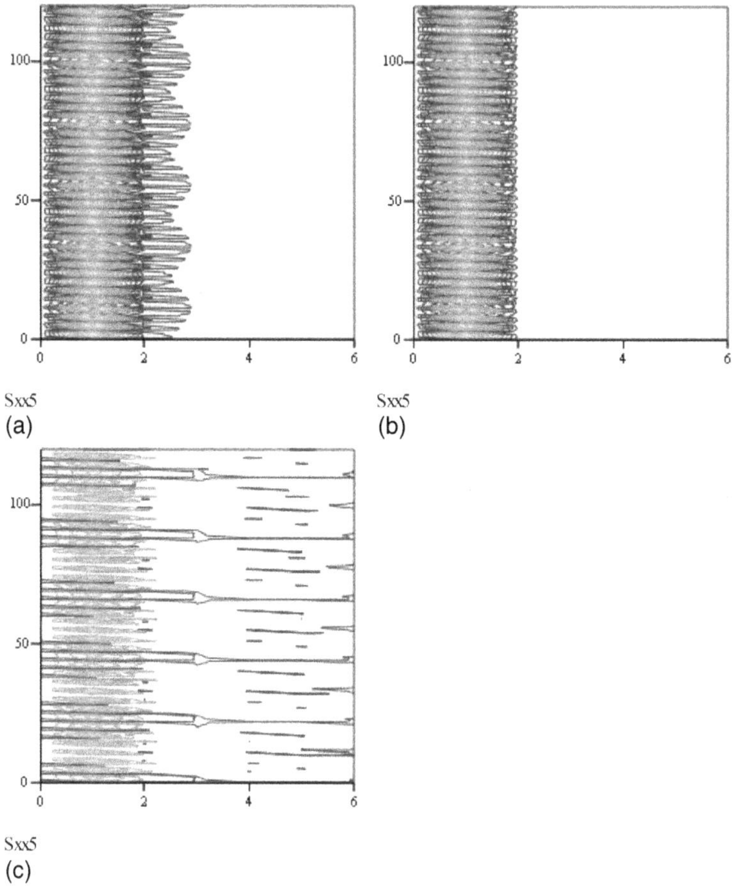


FIGURE 4.19 Plots of $S_{xx5}(t, \omega)$ with $m = 1$, $c = 0$, and $k = 36$ when writing $S_{xx5}(t, \omega) = \text{Sxx5}$ for $\omega = 0, \ldots, 6$ and $t = 0, \ldots, 120$. (a). $S_{xx5}(t, \omega)$ when $H = 0.75$ and $\lambda = 0.2$. (b). $S_{xx5}(t, \omega)$ when $H = 0.75$ and $\lambda = 0.8$. (c). $S_{xx5}(t, \omega)$ when $H = 0.35$ and $\lambda = 0.2$.

**Theorem 4.11 (PSD response VI)**

Let $S_{xx6}(t, \omega)$ be the PSD of $x_6(t)$. Then,

$$S_{xx6}(t,\omega) = \frac{1}{k^2} \frac{\dfrac{V_H}{|\omega|^{2H+1}}\left(1 - 2^{1-2H}\cos 2\omega t\right)}{\left[\omega^{\lambda}\cos\dfrac{\lambda\pi}{2} + \gamma^2\left(\omega^{\alpha-2}\cos\dfrac{\alpha\pi}{2} + 2\varsigma\omega_n\omega^{\beta-2}\cos\dfrac{\beta\pi}{2}\right)\right]^2 + \gamma^2\left(\omega^{\alpha-1}\sin\dfrac{\alpha\pi}{2} + 2\zeta\omega_n\omega^{\beta-1}\sin\dfrac{\beta\pi}{2} + \omega_n^2\omega^{\lambda-1}\sin\dfrac{\lambda\pi}{2}\right)^2}. \tag{4.38}$$

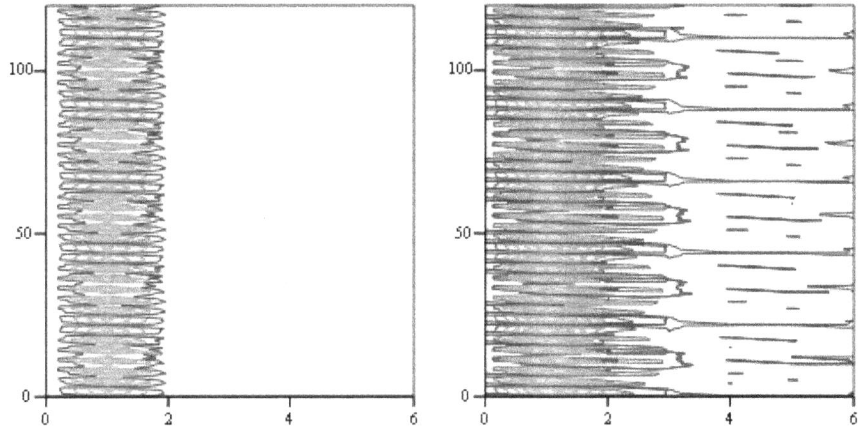


Sfx5
(a)

Sfx5
(b)

FIGURE 4.20 Plots of $|S_{\mathrm{fx5}}(t, \omega)|$ with $m = 1$, $c = 0$, and $k = 36$ when writing $|S_{\mathrm{fx5}}(t, \omega)| = \mathrm{Sfx5}$ for $\omega = 0, \ldots, 6$ and $t = 0, \ldots, 120$. (a). $|S_{\mathrm{fx5}}(t, \omega)|$ when $H = 0.75$ and $\lambda = 0.2$. (b). $|S_{\mathrm{fx5}}(t, \omega)|$ when $H = 0.35$ and $\lambda = 0.2$.

*Proof.* Following Chapter 7 in Volume I or Li [16, 17] or Li [19], we have $H_6(\omega)$ expressed by

$$H_6(\omega) = \frac{1}{k\left[\begin{array}{l} \omega^{\lambda}\cos\frac{\lambda\pi}{2} + \gamma^2\left(\omega^{\alpha-2}\cos\frac{\alpha\pi}{2} + 2\varsigma\omega_n\omega^{\beta-2}\cos\frac{\beta\pi}{2}\right) \\ +i\gamma\left(\omega^{\alpha-1}\sin\frac{\alpha\pi}{2} + 2\zeta\omega_n\omega^{\beta-1}\sin\frac{\beta\pi}{2} + \omega_n^2\omega^{\lambda-1}\sin\frac{\lambda\pi}{2}\right) \end{array}\right]}. \tag{4.39}$$

Due to $S_{xx6}(t, \omega) = S_{ff}(t, \omega)|H_6(\omega)|^2$ and (4.39), we have (4.38). The proof ends.

**Theorem 4.12 (cross-PSD response VI)**

Denote by $S_{fx6}(t, \omega)$ the cross-PSD between $f(t)$ and $x_6(t)$. Then,

$$S_{fx6}(t,\omega) = \frac{\frac{V_H}{|\omega|^{2H+1}}\left(1 - 2^{1-2H}\cos 2\omega t\right)}{k\left[\begin{array}{l} \omega^{\lambda}\cos\frac{\lambda\pi}{2} + \gamma^2\left(\omega^{\alpha-2}\cos\frac{\alpha\pi}{2} + 2\varsigma\omega_n\omega^{\beta-2}\cos\frac{\beta\pi}{2}\right) \\ +i\gamma\left(\omega^{\alpha-1}\sin\frac{\alpha\pi}{2} + 2\zeta\omega_n\omega^{\beta-1}\sin\frac{\beta\pi}{2} + \omega_n^2\omega^{\lambda-1}\sin\frac{\lambda\pi}{2}\right) \end{array}\right]}. \tag{4.40}$$

*Proof.* Performing the operation of $S_{fx6}(t, \omega) = S_{ff}(t, \omega)H_6(\omega)$ with the consideration of (4.39) results in (4.40). The proof is finished.

Doing the inverse Fourier transform on both sides of $S_{xx6}(t, \omega) = S_{ff}(t, \omega)$ $|H_6(\omega)|^2$ produces the ACF response

$$r_{xx6}(t, \tau) = r_{ff}(t, \tau) * h_6(\tau) * h_6(-\tau), \tag{4.41}$$

where $h_6(\tau)$ is the impulse response of a class VI fractional vibrator. The cross-correlation response is given by

$$r_{fx6}(t, \tau) = r_{ff}(t, \tau) * h_6(\tau), \tag{4.42}$$

where $r_{fx6}(t, \tau)$ is the cross-correlation between $f(t)$ and $x_6(t)$. In (4.41) and (4.42), following Chapter 7 in Volume I or Li [16, 17] or Li [19],

$$h_6(t) = e^{-\zeta_{eq6}\omega_{eqn6}t} \frac{1}{m_{eq6}\omega_{eqd6}} \sin \omega_{eqd6} t u(t). \tag{4.43}$$

Refer to Chapter 7 in Volume I or Li [16, 17] or Li [19] for the expressions of $\zeta_{eq6}$, $m_{eq6}$, $\omega_{eqn6}$, and $\omega_{eqd6}$ in (4.43).

### 4.7.2 Effect of $(\alpha, \beta, \lambda)$ on Responses

Figure 4.21 illustrates some plots of $|H_6(\omega)|^2$. Figure 4.22 shows some plots of response PSD $S_{xx6}(t, \omega)$. Figure 4.23 demonstrates some plots of

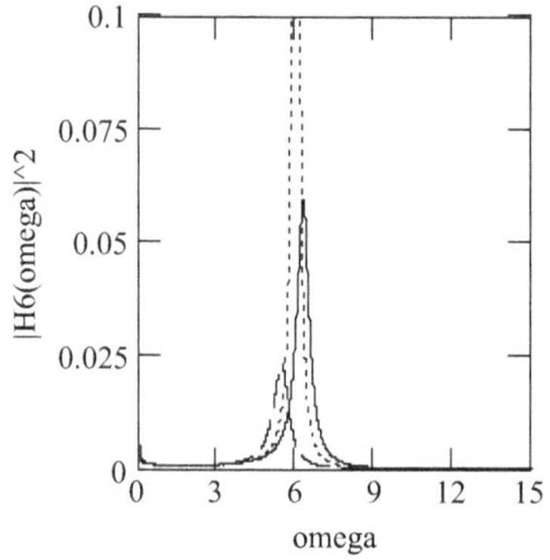


FIGURE 4.21 Plots of $|H_6(\omega)|^2$ when $m = 1$, $c = 0.1$, and $k = 36$, for $(\alpha, \beta, \lambda) =$ (2.1, 0.8, 0.15) (solid), (2.2, 1.2, 0.20) (dot), (2.3, 1.6, 0.25) (dash).

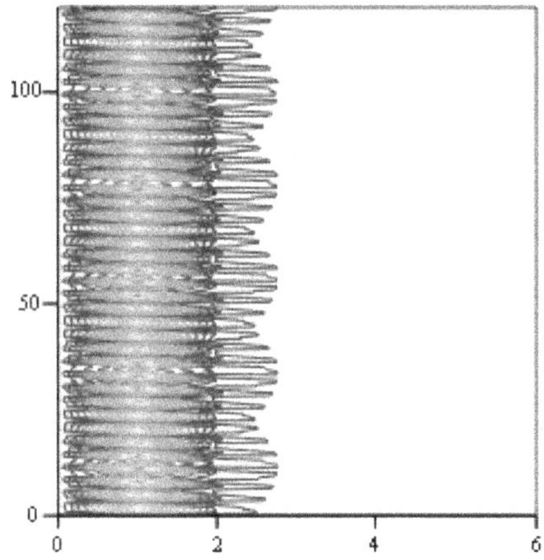


Sxx6

(a)

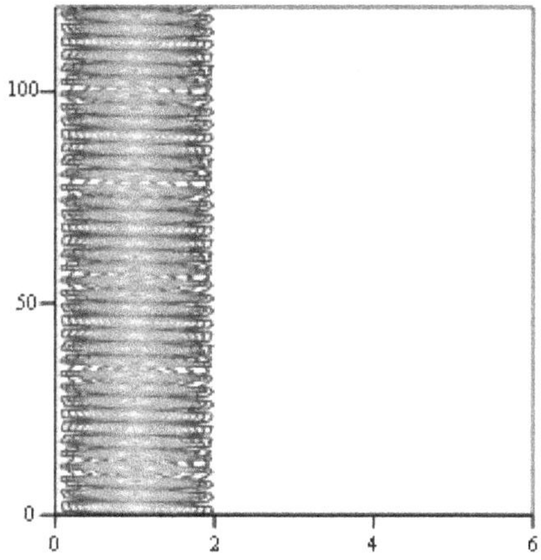


Sxx6

(b)

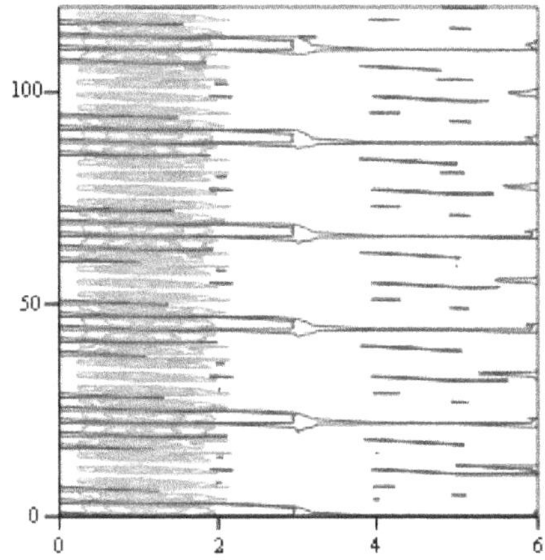


Sxx6

(c)

FIGURE 4.22 Plots of $S_{xx6}(t, \omega)$ with $m = 1$, $c = 0.1$, and $k = 36$ when writing $S_{xx6}(t, \omega) =$ Sxx6. (a). $S_{xx6}(t, \omega)$ when $H = 0.75$ and $(\alpha, \beta, \lambda) = (1.8, 0.8, 0.2)$ when writing $S_{xx6}(t, \omega) =$ Sxx6 for $\omega = 0, \ldots, 6$ and $t = 0, \ldots, 120$. (b). $S_{xx6}(t, \omega)$ when $H = 0.75$ and $(\alpha, \beta, \lambda) = (1.8, 1.8, 0.8)$. (c). $S_{xx6}(t, \omega)$ when $H = 0.35$ and $(\alpha, \beta, \lambda) = (1.8, 0.8, 0.2)$.

$|S_{fx6}(t, \omega)|$. Figures 4.22 and 4.23 imply that the effect of $(\alpha, \beta, \lambda)$ on the responses to class VI fractional vibrators under the excitation of fBm is significant.

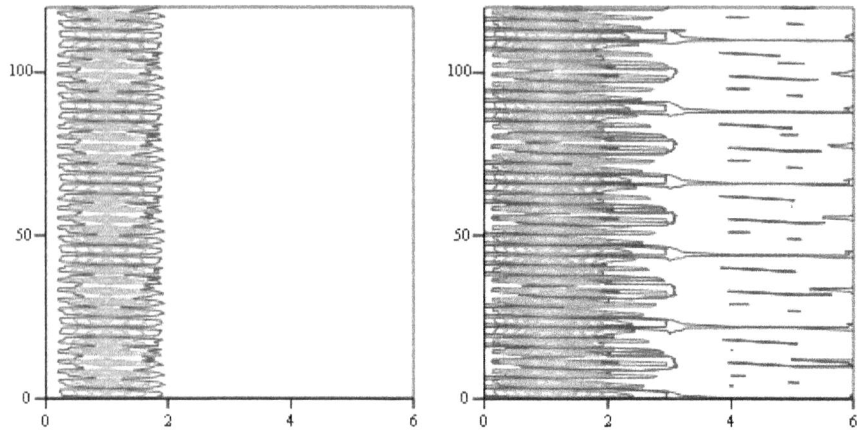


Sfx6
(a)

Sfx6
(b)

FIGURE 4.23 Plots of $|S_{fx6}(t,\omega)|$ with $m=1, c=0.1$, and $k=36$ when writing $|S_{fx6}(t,\omega)| =$ Sfx6 for $\omega = 0, \ldots, 6$ and $t = 0, \ldots, 120$. (a). $|S_{fx6}(t, \omega)|$ when $H = 0.75$ and $(\alpha, \beta, \lambda) = (1.8, 0.8, 0.2)$. (b). $|S_{fx6}(t, \omega)|$ when $H = 0.35$ and $(\alpha, \beta, \lambda) = (1.8, 0.8, 0.2)$.

## 4.8 RESPONSES OF CLASS VII FRACTIONAL VIBRATORS DRIVEN BY FBM

### 4.8.1 Computations

The motion equation of a class VII fractional vibrator is given by (4.44) in the form

$$m\frac{d^2x_7(t)}{dt^2}+c\frac{d^\beta x_7(t)}{dt^\beta}+k\frac{d^\lambda x_7(t)}{dt^\lambda}=f(t), \tag{4.44}$$

where $x_7(t)$ is the response of a class VII fractional vibrator.

**Theorem 4.13 (PSD response VII)**

Denote by $S_{xx7}(t, \omega)$ the PSD of $x_7(t)$. Then,

$$S_{xx7}(t,\omega)=\frac{1}{k^2}\frac{\frac{V_H}{|\omega|^{2H+1}}\left(1-2^{1-2H}\cos 2\omega t\right)}{\left[\omega^\lambda\cos\frac{\lambda\pi}{2}-\gamma\left(1-2\varsigma\omega_n\omega^{\beta-2}\cos\frac{\beta\pi}{2}\right)\right]^2+\gamma^2\left(2\varsigma\omega^{\beta-1}\sin\frac{\beta\pi}{2}+\omega_n\omega^{\lambda-1}\sin\frac{\lambda\pi}{2}\right)^2}. \tag{4.45}$$

*Proof.* Doing the operation of $S_{xx7}(t, \omega) = S_{ff}(t, \omega)|H_7(\omega)|^2$, where $H_7(\omega)$ is expressed by (Chapter 7 in Volume I or Li [19])

$$H_7(\omega) = \frac{1}{k\left[\begin{array}{l}\omega^{\lambda}\cos\dfrac{\lambda\pi}{2} - \gamma\left(1 - 2\varsigma\omega_n\omega^{\beta-2}\cos\dfrac{\beta\pi}{2}\right)\\ +i\gamma\left(2\varsigma\omega^{\beta-1}\sin\dfrac{\beta\pi}{2} + \omega_n\omega^{\lambda-1}\sin\dfrac{\lambda\pi}{2}\right)\end{array}\right]}, \tag{4.46}$$

results in (4.45). The proof completes.

**Theorem 4.14 (cross-PSD response VII)**

Let $S_{fx7}(t, \omega)$ be the cross-PSD between $f(t)$ and $x_7(t)$. Then,

$$S_{fx7}(t,\omega) = \frac{\dfrac{V_H}{|\omega|^{2H+1}}\left(1 - 2^{1-2H}\cos 2\omega t\right)}{k\left[\begin{array}{l}\omega^{\lambda}\cos\dfrac{\lambda\pi}{2} - \gamma\left(1 - 2\varsigma\omega_n\omega^{\beta-2}\cos\dfrac{\beta\pi}{2}\right)\\ +i\gamma\left(2\varsigma\omega^{\beta-1}\sin\dfrac{\beta\pi}{2} + \omega_n\omega^{\lambda-1}\sin\dfrac{\lambda\pi}{2}\right)\end{array}\right]}. \tag{4.47}$$

*Proof.* Doing the operation of $S_{fx7}(\omega) = S_{ff}(\omega)H_7(\omega)$ and taking into account (4.46) yields (4.47). The proof ends.

Doing the inverse Fourier transform on both sides of $S_{xx7}(t, \omega) = S_{ff}(t, \omega)$ $|H_7(\omega)|^2$ produces the ACF response given by

$$r_{xx7}(t, \tau) = r_{ff}(t, \tau) * h_7(\tau) * h_7(-\tau), \tag{4.48}$$

where $h_7(\tau)$ is the impulse response of a class VII fractional vibrator. Similarly, performing the inverse Fourier transform on both sides of $S_{fx7}(\omega) = S_{ff}(\omega)H_7(\omega)$ yields the cross-correlation response

$$r_{fx7}(t, \tau) = r_{ff}(t, \tau) * h_7(\tau). \tag{4.49}$$

In (4.48) and (4.49), $h_7(t)$ is (Chapter 7 in Volume I or Li [19])

$$h_7(t) = e^{-\varsigma_{\text{eq7}}\omega_{\text{eqn7}}t}\frac{1}{m_{\text{eq7}}\omega_{\text{eqd7}}}\sin\omega_{\text{eqd7}}t, \quad t \geq 0. \tag{4.50}$$

Refer to Chapter 7 in Volume I or Li [19] for the expressions of $\zeta_{eq7}$, $m_{eq7}$, $\omega_{eqn7}$, and $\omega_{eqd7}$ in (4.50).

### 4.8.2 Effect of $(\beta, \lambda)$ on Responses

Figure 4.24 indicates some plots of $|H_7(\omega)|^2$, Figure 4.25 shows some plots of response PSD $S_{xx7}(t, \omega)$, and Figure 4.26 is for some plots of $|S_{fx7}(t, \omega)|$.

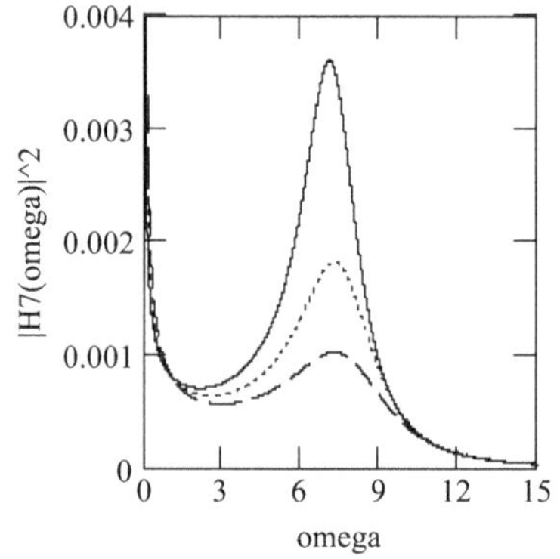


FIGURE 4.24 Plots of $|H_7(\omega)|^2$ (log) when $m = 1$, $c = 0.1$, and $k = 36$, for $(\beta, \lambda) =$ (0.5, 0.20) (solid), (1.0, 0.25) (dot), (1.5, 0.30) (dash).

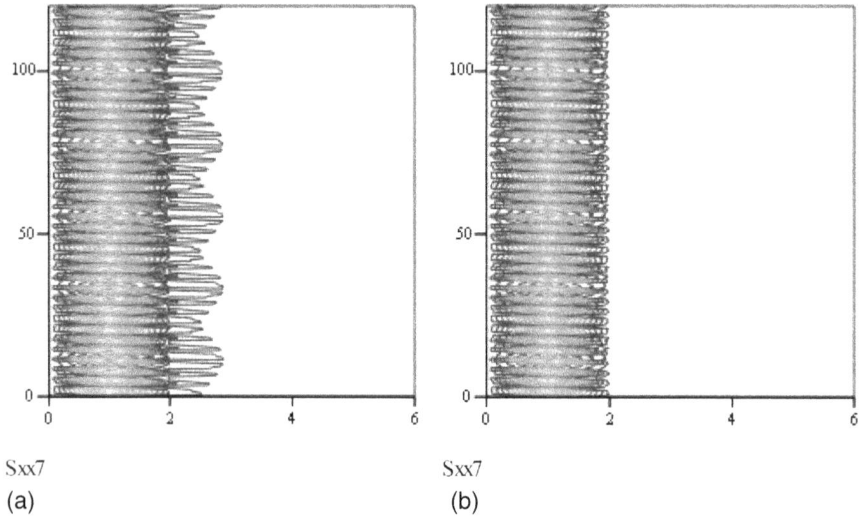


FIGURE 4.25 Plots of $S_{xx7}(t, \omega)$ with $m = 1$, $c = 0.1$, and $k = 36$ when writing $S_{xx7}(t, \omega) =$ Sxx7 for $\omega = 0, \ldots, 6$ and $t = 0, \ldots, 120$. (a). $S_{xx7}(t, \omega)$ when $H = 0.75$ and $(\beta, \lambda) = (0.8, 0.2)$. (b). $S_{xx7}(t, \omega)$ when $H = 0.75$ and $(\beta, \lambda) =$ (0.8, 0.8). (c). $S_{xx7}(t, \omega)$ when $H = 0.35$ and $(\beta, \lambda) = (0.8, 0.8)$.

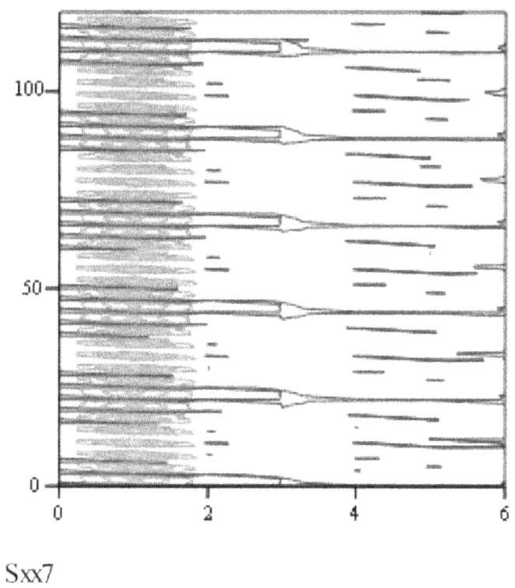


(c)

FIGURE 4.25 (Continued)

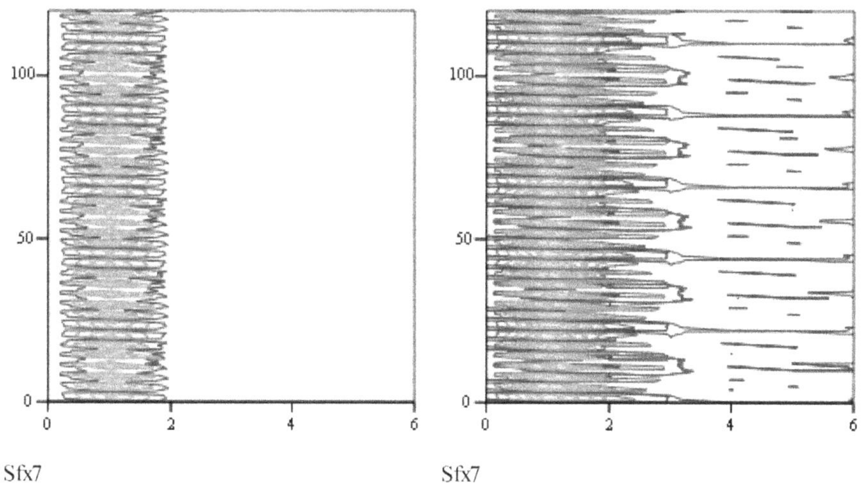


(a) (b)

FIGURE 4.26 Plots of $|S_{fx7}(t, \omega)|$ with $m = 1$, $c = 0.1$, and $k = 36$ when writing $|S_{fx7}(t, \omega)| = \text{Sxx7}$ for $\omega = 0, \ldots, 6$ and $t = 0, \ldots, 120$. (a). $|S_{fx7}(t, \omega)|$ for $H = 0.75$ and $(\beta, \lambda) = (0.8, 0.2)$. (b). $|S_{fx7}(t, \omega)|$ for $H = 0.35$ and $(\beta, \lambda) = (0.8, 0.2)$. (c). $|S_{fx7}(t, \omega)|$ for $H = 0.75$ and $(\beta, \lambda) = (1.8, 1.8)$. (d). $|S_{fx7}(t, \omega)|$ for $H = 0.35$ and $(\beta, \lambda) = (1.8, 1.8)$.

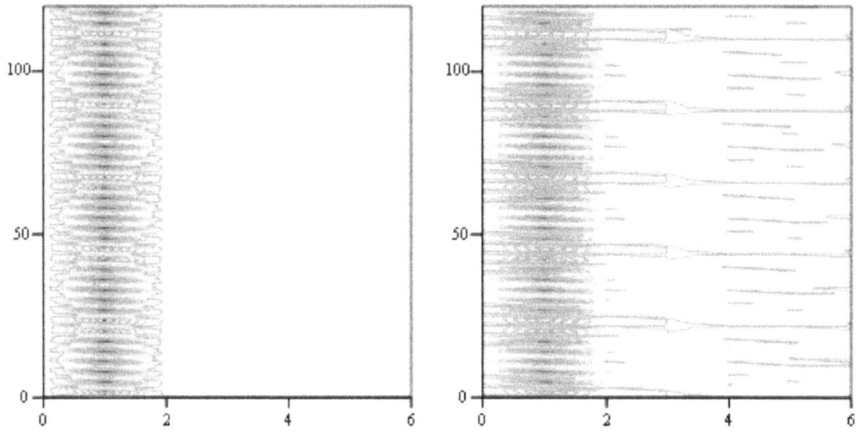


Sfx7
(c)

Sfx7
(d)

FIGURE 4.26 (Continued)

Figures 4.25 and 4.26 exhibit that the effect of ($\beta$, $\lambda$) on the responses to class VII fractional vibration systems driven by fBm is significant.

## 4.9 SUMMARY

The analytic expressions of PSD and cross-PSD responses to seven classes of fractional vibrators driven by fBm have been given in Theorems 4.1–4.14, respectively. The noticeable effects of fractional orders of vibration systems on the responses have been illustrated. The responses are of long-range dependence for $0 < H < 0.5$ and $0.5 < H < 1$ because $S_{xxi}(t, 0) = \infty$ when $0 < H < 0.5$ and $0.5 < H < 1$ for $i = 1, \ldots, 7$.

## 4.10 EXERCISES

4.1. Let

$$S_{ff}(t,\omega) = \frac{V_H}{|\omega|^{2H+1}}\left(1 - 2^{1-2H}\cos 2\omega t\right),$$

where $0 < H < 1$ and $V_H = \Gamma(1-2H)\dfrac{\cos \pi H}{\pi H}$. Find the inverse Fourier transform of $S_{ff}(t, \omega)$.

4.2. Following Chapter 7 in Volume I, we denote

$$h_5(t) = e^{-\varsigma_{\mathrm{eq5}}\omega_{\mathrm{eqn5}}t}\frac{1}{m\omega_{\mathrm{eqd5}}}\sin\omega_{\mathrm{eqd5}}tu(t)$$

$$= e^{-\frac{k\omega^{\lambda-1}\sin\frac{\lambda\pi}{2}}{2\sqrt{mk\omega^{\lambda}\cos\frac{\lambda\pi}{2}}}\sqrt{\omega^{\lambda}\cos\frac{\lambda\pi}{2}}\omega_n t}\frac{1}{m\omega_{\mathrm{eqd5}}}\sin\omega_{\mathrm{eqd5}}tu(t).$$

Find $h_5(\tau)*h_5(-\tau)$.

4.3. Find $r_{xx5}(t,\tau) = r_{ff}(t,\tau)*h_5(\tau)*h_5(-\tau)$, where $r_{ff}(\tau) = \mathrm{F}^{-1}[S_{ff}(\omega)]$ in Exercise 4.1.

4.4. Find $r_{fx5}(t,\tau) = r_{ff}(t,\tau)*h_5(\tau)$, where $r_{ff}(\tau) = \mathrm{F}^{-1}[S_{ff}(\omega)]$ in Exercise 4.1.

4.5. Find the condition for $\int_{-\infty}^{\infty} r_{xx5}(t,\tau)d\tau = \infty$.

4.6. Find the condition for $S_{ff}(t,0) = \infty$.

## REFERENCES


1. J. He, Z. Xing, and Q. Guo, Spectral collocation method for stochastic differential equations driven by fractional Brownian motion, *Fluctuation and Noise Letters*, 22(3): 2023, 2350019.
2. J. He and Q. Guo, An explicit method for the self-interacting diffusion driven by fractional Brownian motion under global Lipschitz conditions, *Applied Mathematics Letters*, 134: 2022, 108379.
3. X. Liu, Strong approximation for fractional wave equation forced by fractional Brownian motion with Hurst parameter $H \in (0, 0.5)$, *Journal of Computational and Applied Mathematics*, 432: 2023, 115285.
4. N. H. Tuan, T. Caraballo, and T. N. Thach, New results for stochastic fractional pseudo-parabolic equations with delays driven by fractional Brownian motion, *Stochastic Processes and Their Applications*, 161: 2023, 24–67.
5. N. H. Tuan, M. Foondun, T. N. Thach, and R. Wang, On backward problems for stochastic fractional reaction equations with standard and fractional Brownian motion, *Bulletin des Sciences Mathématiques*, 179: 2022, 103158.
6. N. Sharma, D. Selvamuthu, and S. Natarajan, Variable annuities valuation under a mixed fractional Brownian motion environment with jumps considering mortality risk, *Applied Stochastic Models in Business and Industry*, 38(6): 2022, 1019–1038.
7. X. Fan, X. Huang, Y. Suo, and C. Yuan, Distribution dependent SDEs driven by fractional Brownian motions, *Stochastic Processes and their Applications*, 151: 2022, 23–67.

8. S.-Q. Zhang and C. Yuan, Stochastic differential equations driven by fractional Brownian motion with locally Lipschitz drift and their implicit Euler approximation, *Proceedings of the Royal Society of Edinburgh Section A: Mathematics*, 151(4): 2021, 1278–1304.
9. X. Sun, R. Guo, and M. Li, Some properties of bifractional Bessel processes driven by bifractional Brownian motion, *Mathematical Problems in Engineering*, 2020, 7037602, 13.
10. A. Shahnazi-Pour, B. Parsa Moghaddam, and A. Babaei, Numerical simulation of the Hurst index of solutions of fractional stochastic dynamical systems driven by fractional Brownian motion, *Journal of Computational and Applied Mathematics*, 386: 2021, 113210.
11. H. Araya, J. A. León, and S. Torres, Numerical scheme for stochastic differential equations driven by fractional Brownian motion with $1/4 < H < 1/2$, *Journal of Theoretical Probability*, 33(3): 2020, 1211–1237.
12. J. Gairing, P. Imkeller, R. Shevchenko, and C. Tudor, Hurst index estimation in stochastic differential equations driven by fractional Brownian motion, *Journal of Theoretical Probability*, 33(3): 2020, 1691–1714.
13. P. Xu, J. Huang, and C. Zeng, Ergodicity of stochastic Rabinovich systems driven by fractional Brownian motion, *Physica A*, 546: 2020, 122955.
14. M. H. Heydari, Z. Avazzadeh, and M. R. Mahmoudi, Chebyshev cardinal wavelets for nonlinear stochastic differential equations driven with variable-order fractional Brownian motion, *Chaos, Solitons & Fractals*, 124: 2019, 105–124.
15. M. H. Heydari, M. R. Mahmoudi, A. Shakiba, and Z. Avazzadeh, Chebyshev cardinal wavelets and their application in solving nonlinear stochastic differential equations with fractional Brownian motion, *Communications in Nonlinear Science and Numerical Simulation*, 64: 2018, 98–121.
16. M. Li, *Fractional Vibrations with Applications to Euler-Bernoulli Beams*, CRC Press, Boca Raton, 2023.
17. M. Li, *Theory of Fractional Engineering Vibrations*, Walter de Gruyter, Berlin/Boston, 2021.
18. M. Li, Three classes of fractional oscillators, *Symmetry-Basel*, 10(2): 2018, 91.
19. M. Li, Analytic theory of seven classes of fractional vibrations based on elementary functions: A tutorial review, *Symmetry*, 16(9): 2024, 1202.
20. M. Li, PSD and cross PSD of responses of seven classes of fractional vibrations driven by fGn, fBm, fractional OU process, and von Kármán process, *Symmetry*, 16(5): 2024, 635.

CHAPTER 5

# Responses of Fractional Vibrations Driven by Fractional Ornstein-Uhlenbeck Processes

THIS CHAPTER CONTRIBUTES THE analytic expressions of power spectrum density (PSD) and cross-PSD responses to seven classes of fractional vibrators driven by the fractional Ornstein-Uhlenbeck (OU) processes. Results exhibit that the fractional orders of vibration systems as well as the order of fractional OU processes have effects on responses. The responses of seven classes of fractional vibrators driven by fractional OU processes are of short-range dependence.

## 5.1 BACKGROUND

The Langevin equation and the Ornstein-Uhlenbeck (OU) process, which is the solution to the Langevin equation under the excitation of white noise, remain the research interests in various fields (Pavliotis [1], Coffey et al. [2], and Eliazar and Shlesinger [3]). Naturally, fractional Langevin equations attract researchers (West et al. [4], Eab and Lim [5, 6], Lim et al. [7, 8], Li et al. [9]). Accordingly, fractional OU processes have been paid attention to; see, for example, Coffey et al. [2], Shao [10], Cheridito et al. [11], Magdziarz [12], Gehringer and Li [13], Patel and Sharma [14], Lim

DOI: 10.1201/9781003657903-5

and Muniandy [15], simply mentioning a few. One thing relating to marine structure vibrations is the von Kármán process, which is in fact a kind of fractional OU processes (Li [16]).

In the field, analytic expressions of the responses to seven classes of fractional vibrators driven by the fractional OU processes are rarely reported. In this chapter, we contribute the closed-form analytic expressions of the PSD and cross-PSD responses to seven classes of fractional vibrators under the excitation of fractional OU processes. The present results exhibit that there is considerable effect of fractional order, $\alpha$, or $\beta$, or $\lambda$, on responses.

The rest of the chapter is organized as follows. In Sections 5.2–5.8, we address the analytic expressions of the PSD and cross-PSD responses to the fractional vibration systems from classes I to VII that are driven by the fractional OU processes. The summary is given in Section 5.9.

## 5.2 RESPONSES OF CLASS I FRACTIONAL VIBRATORS DRIVEN BY FRACTIONAL OU PROCESSES

### 5.2.1 Computations

Consider the motion equation of a class I fractional vibrator in the form

$$m\frac{d^{\alpha}x_1(t)}{dt^{\alpha}}+k\frac{dx_1(t)}{dt}=f(t). \tag{5.1}$$

In (5.1), $1<\alpha<3$, $x_1(t)$ is the response, $f(t)$ is the excitation, $m$ and $k$ are the primary mass and stiffness, respectively.

Let $f(t)$ be a fractional OU process of order $b$ in what follows. Denote by $S_{ff}(\omega)$ the PSD of $f(t)$. Following Chapter 3 in Volume I, or [3–9], one has

$$S_{ff}(\omega)=\frac{1}{\left(\lambda^2+\omega^2\right)^b}. \tag{5.2}$$

In (5.2), $b>0$, $\lambda>0$. Without the generality losing, we set $\lambda=1$ in what follows. Figure 5.1 illustrates some plots of $S_{ff}(\omega)$.

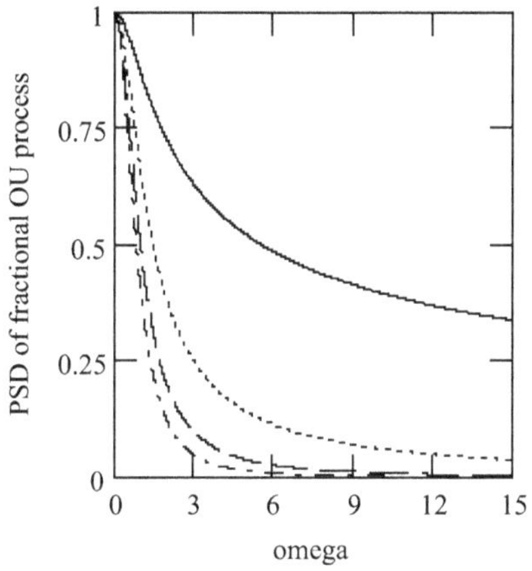


FIGURE 5.1 Plots of PSD of fractional OU processes for $b = 0.2$ (solid), 0.6 (dot), 1 (dash), 1.3 (dash dot).

**Theorem 5.1 (PSD response I)**

Denote by $S_{xx1}(\omega)$ the PSD of $x_1(t)$. Then,

$$S_{xx1}(\omega) = \frac{\dfrac{1}{\left(\lambda^2 + \omega^2\right)^b}}{k^2\left[\left(1 - \dfrac{\omega^\alpha}{\omega_n^2}\left|\cos\dfrac{\alpha\pi}{2}\right|\right)^2 + \left(\dfrac{\omega^\alpha}{\omega_n^2}\sin\dfrac{\alpha\pi}{2}\right)^2\right]}, \tag{5.3}$$

where $\omega_n^2 = \dfrac{k}{m}$.

*Proof.* Note that $S_{xx1}(\omega) = S_{ff}(\omega)|H_1(\omega)|^2$. Based on Chapter 7 in Volume I or Li [16, 17], $H_1(\omega)$ is given by

$$H_1(\omega) = \frac{1}{k\left(1 - \dfrac{\omega^\alpha}{\omega_n^2}\left|\cos\dfrac{\alpha\pi}{2}\right| + i\dfrac{\omega^\alpha}{\omega_n^2}\sin\dfrac{\alpha\pi}{2}\right)}. \tag{5.4}$$

Thus, we have (5.3). The proof is finished.

**Theorem 5.2 (cross-PSD response I)**

Let $S_{fx1}(\omega)$ be the cross-PSD between $f(t)$ and $x_1(t)$. Then,

$$S_{fx1}(\omega) = \frac{\frac{1}{\left(\lambda^2+\omega^2\right)^b}}{k\left(1-\frac{\omega^\alpha}{\omega_n^2}\left|\cos\frac{\alpha\pi}{2}\right|+i\frac{\omega^\alpha}{\omega_n^2}\sin\frac{\alpha\pi}{2}\right)}. \tag{5.5}$$

*Proof.* Taking into account $S_{fx1}(\omega) = S_{ff}(\omega)H_1(\omega)$ and (5.4), we have (5.5). The proof ends.

Let $r_{xx1}(\tau)$ be the autocorrelation function (ACF) of $x_1(t)$. Let $r_{ff}(\tau)$ be the ACF of $f(t)$. Let $r_{fx1}(\tau)$ be the cross-correlation between the excitation $f(t)$ and the response $x_1(t)$. Denote by $h_1(\tau)$ the impulse response of a class I fractional vibrator. Then, the ACF response $r_{xx1}(\tau)$ is given by

$$r_{xx1}(\tau) = r_{ff}(\tau)*h_1(\tau)*h_1(-\tau), \tag{5.6}$$

and the cross-correlation response $r_{fx1}(\tau)$ is expressed by

$$r_{fx1}(\tau) = r_{ff}(\tau)*h_1(\tau). \tag{5.7}$$

In (5.6) and (5.7) (Chapter 7 in Volume I, Li [16, 17]),

$$h_1(t) = \frac{e^{-\frac{\omega\sin\frac{\alpha\pi}{2}}{2\left|\cos\frac{\alpha\pi}{2}\right|}t}\sin\left(\frac{\omega_n}{\sqrt{\omega^{\alpha-2}\left|\cos\frac{\alpha\pi}{2}\right|}}\sqrt{1-\frac{\omega^{2\alpha}\sin^2\frac{\alpha\pi}{2}}{4\omega_n^2\left|\cos\frac{\alpha\pi}{2}\right|}}\,t\right)}{m\omega_n\sqrt{\omega^{\alpha-2}\left|\cos\frac{\alpha\pi}{2}\right|}\sqrt{1-\frac{\omega^{2\alpha}\sin^2\frac{\alpha\pi}{2}}{4\omega_n^2\left|\cos\frac{\alpha\pi}{2}\right|}}}u(t). \tag{5.8}$$

In (5.8), $u(t)$ is the unit step function.

### 5.2.2 Effect of $\alpha$ on Responses

Figure 5.2 indicates some plots of $|H_1(\omega)|^2$. Some plots of $S_{xx1}(\omega)$ are shown in Figure 5.3. When $\alpha = 2$, a class I fractional vibrator reduces to be a

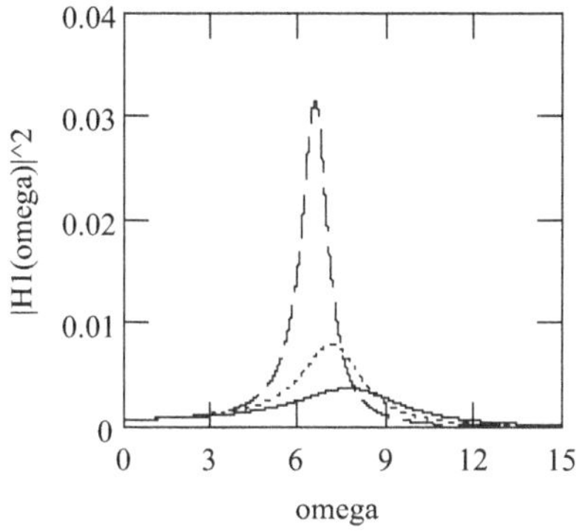


FIGURE 5.2 Plots of $|H_1(\omega)|^2$ with $\alpha = 1.7$ (solid), 1.8 (dot), 1.9 (dash) when $m = 1$ and $k = 36$.

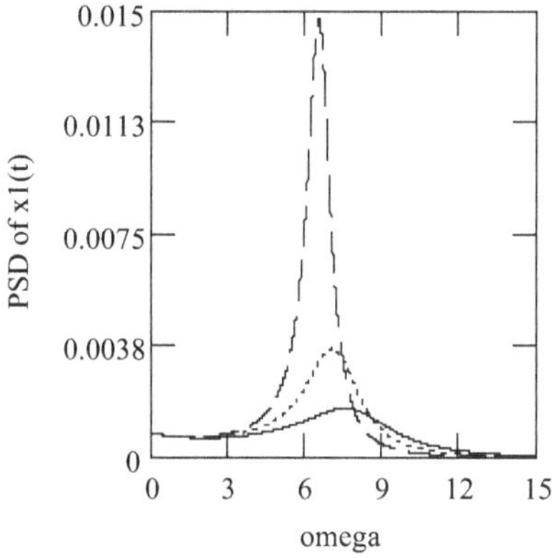


FIGURE 5.3 Plots of $S_{xx1}(\omega)$ for $m = 1$ and $k = 36$ when $(\alpha, b) = (1.7, 0.2)$ (solid), (1.8, 0.2) (dot), (1.5, 1) (dash), (1.9, 0.2) (dash dot).

conventional damping-free vibrator. Figure 5.4 shows the effect of $b$ on $S_{xx1}(\omega)$. Figure 5.5 illustrates some plots of $S_{fx1}(\omega)$. Figures 5.4 and 5.5 exhibit that the order $\alpha$ of class I fractional vibration systems and the order $b$ of the fractional OU processes have noticeable effects on the responses to class I fractional vibrators. We illustrate some plots of driven fractional OU processes and response ones in Figure 5.6.

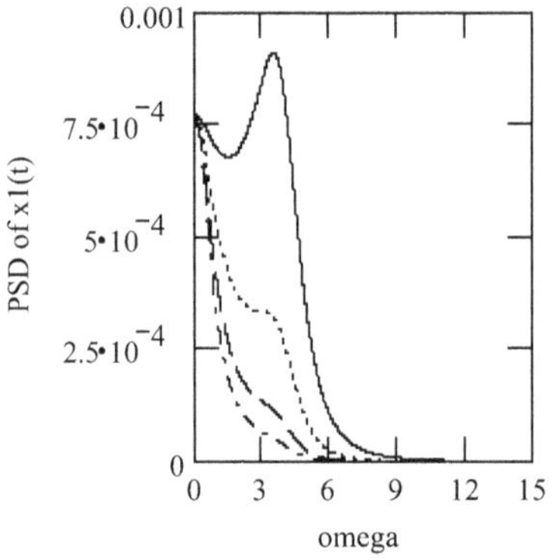


FIGURE 5.4 Observing $b$ effect on $S_{xx1}(\omega)$ with $\alpha = 2.5$ for $b = 0.2$ (solid), 0.6 (dot), 1 (dash), 1.3 (dash dot).

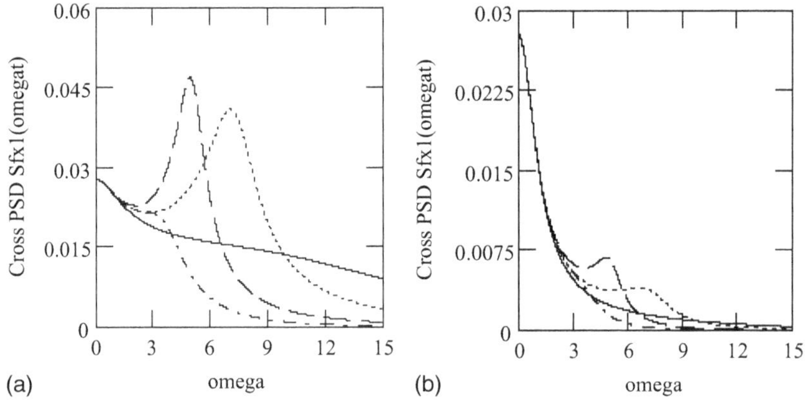


FIGURE 5.5 Illustrations of $|S_{fx1}(\omega)|$. (a). $|S_{fx1}(\omega)|$ for $m = 1$ and $k = 36$ when $(\alpha, b) = (1.4, 0.2)$ (solid), (1.8, 0.2) (dot), (2.2, 0.2) (dash), (2.6, 0.2) (dash dot). (b). $|S_{fx1}(\omega)|$ for $m = 1$ and $k = 36$ when $(\alpha, b) = (1.4, 0.8)$ (solid), (1.8, 0.8) (dot), (2.2, 0.8) (dash), (2.6, 0.8) (dash dot).

## 5.3 RESPONSES OF CLASS II FRACTIONAL VIBRATION SYSTEMS DRIVEN BY FRACTIONAL OU PROCESSES

### 5.3.1 Computation Methods

For a class II fractional vibrator, its motion equation is given by

$$m\frac{d^2 x_2(t)}{dt^2} + c\frac{d^\beta x_2(t)}{dt^\beta} + k\frac{dx_2(t)}{dt} = f(t). \tag{5.9}$$

In (4.9), $0 < \beta < 2$, $x_2(t)$ is the response driven by excitation $f(t)$ and $c$ is the primary damping.

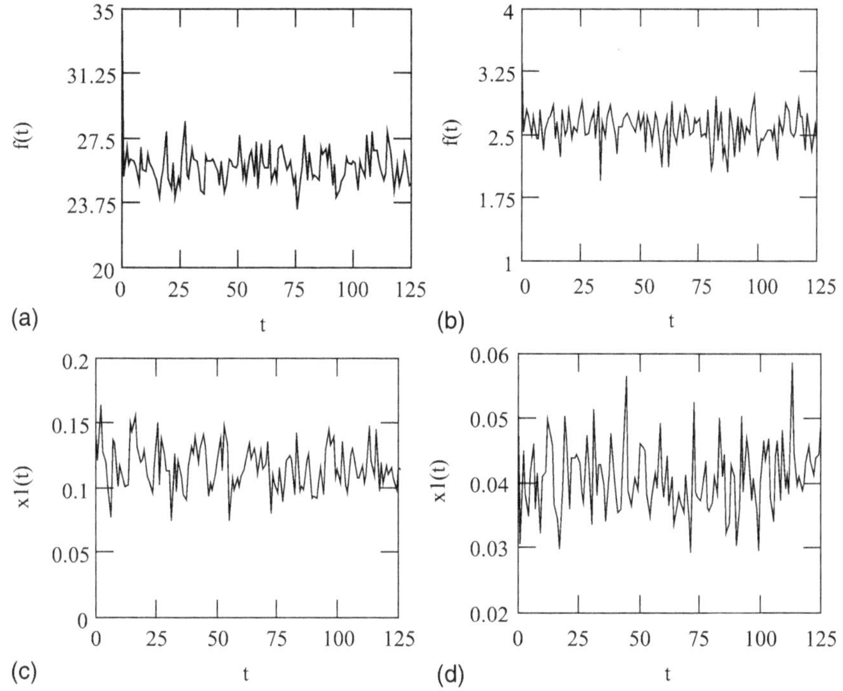


FIGURE 5.6 Plots of driven fractional OU signal and response signal when $m = 1$, $k = 36$. (a). Driven fractional OU signal for $b = 0.2$. (b). Driven fractional OU signal for $b = 0.8$. (c). Response $x_1(t)$ for $(\alpha, b) = (1.8, 0.2)$. (d). Response $x_1(t)$ for $(\alpha, b) = (1.2, 0.8)$.

**Theorem 5.3 (PSD response II)**

Let $S_{xx2}(\omega)$ be the PSD of $x_2(t)$. Then,

$$S_{xx2}(\omega) = \frac{\dfrac{1}{\left(\lambda^2+\omega^2\right)^b}}{k^2\left\{\left[1-\gamma^2\left(1-\dfrac{c}{m}\omega^{\beta-2}\cos\dfrac{\beta\pi}{2}\right)\right]^2+\left(\dfrac{2\varsigma\omega^\beta}{\omega_n}\sin\dfrac{\beta\pi}{2}\right)^2\right\}}, \tag{5.10}$$

where $\gamma = \dfrac{\omega}{\omega_n}$.

*Proof.* Doing the operation of $S_{xx2}(\omega) = S_{ff}(\omega)|H_2(\omega)|^2$, where $H_2(\omega)$ is given by (Chapter 7 in Volume I or Li [16, 17])

$$H_2(\omega)=\frac{1/k}{1-\gamma^2\left(1-\frac{c}{m}\omega^{\beta-2}\cos\frac{\beta\pi}{2}\right)+i\frac{2\varsigma\omega^{\beta}\sin\frac{\beta\pi}{2}}{\omega_n}},\tag{5.11}$$

we have (5.10). This finishes the proof.

**Theorem 5.4 (cross-PSD response II)**

Denote by $S_{fx2}(\omega)$ the cross PSD between $f(t)$ and $x_2(t)$. Then,

$$S_{fx2}(\omega)=\frac{\frac{1}{\left(\lambda^2+\omega^2\right)^b}}{k\left[1-\gamma^2\left(1-2\varsigma\omega_n\omega^{\beta-2}\cos\frac{\beta\pi}{2}\right)+i\frac{2\varsigma\omega^{\beta}}{\omega_n}\sin\frac{\beta\pi}{2}\right]}.\tag{5.12}$$

*Proof.* Because of $S_{fx2}(\omega)=S_{ff}(\omega)H_2(\omega)$ and (5.11), (5.12) holds. The proof ends.

Let $r_{xx2}(\tau)$ be the ACF of $x_2(t)$. Let $r_{fx2}(\tau)$ be the cross-correlation between the excitation $f(t)$ and the response $x_2(t)$. Denote by $h_2(\tau)$ the impulse response of a class II fractional vibrator. According to the convolution theory, one has the ACF response $r_{xx2}(\tau)$ expressed by

$$r_{xx2}(\tau)=r_{ff}(\tau)*h_2(\tau)*h_2(-\tau),\tag{5.13}$$

and the cross-correlation response $r_{fx2}(\tau)$ given by

$$r_{fx2}(\tau)=r_{ff}(\tau)*h_2(\tau).\tag{5.14}$$

In (5.13) and (5.14), according to Chapter 7 in Volume I or Li [16, 17], $h_2(t)$ is expressed by

$$h_2(t)=\frac{e^{-\frac{\varsigma\omega_n\omega^{\beta-1}\sin\frac{\beta\pi}{2}}{1-\frac{c}{m}\omega^{\beta-2}\cos\frac{\beta\pi}{2}}t}\sin\frac{\omega_n\sqrt{1-\frac{\varsigma^2\omega^{2(\beta-1)}\sin^2\frac{\beta\pi}{2}}{1-\frac{c}{m}\omega^{\beta-2}\cos\frac{\beta\pi}{2}}}}{\sqrt{1-\frac{c}{m}\omega^{\beta-2}\cos\frac{\beta\pi}{2}}}t}{\omega_n m\sqrt{1-\frac{c}{m}\omega^{\beta-2}\cos\frac{\beta\pi}{2}}\sqrt{1-\frac{\varsigma^2\omega^{2(\beta-1)}\sin^2\frac{\beta\pi}{2}}{1-\frac{c}{m}\omega^{\beta-2}\cos\frac{\beta\pi}{2}}}}u(t).\tag{5.15}$$

### 5.3.2 Effect of $\beta$ on Responses

Some plots of $|H_2(\omega)|^2$ are indicated in Figure 5.7. Figure 5.8 shows some plots of response PSD $S_{xx2}(\omega)$. Figure 5.9 shows that the order $b$ of the fractional OU processes does not obviously affect $S_{xx2}(\omega)$. Figure 5.10 illustrates some plots of $|S_{fx2}(\omega)|$. Figures 5.8 and 5.10 show that there is effect of $\beta$ on the responses to class II fractional vibrators driven by fractional OU processes. Figure 5.11 exhibits the effect of $\beta$ on the fluctuation range of $x_2(t)$.

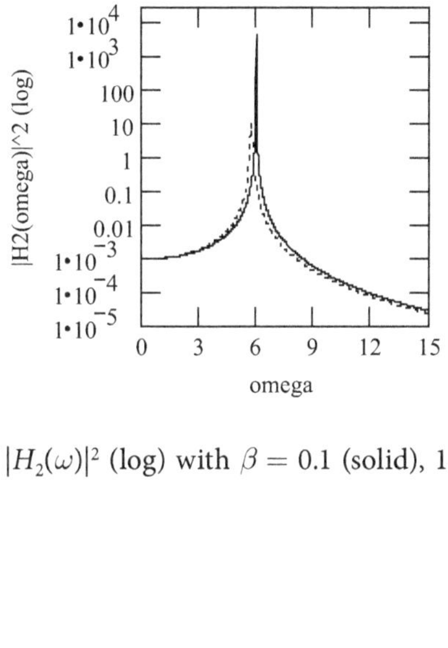


FIGURE 5.7 Plots of $|H_2(\omega)|^2$ (log) with $\beta = 0.1$ (solid), 1.9 (dot), when $m = 1$, $c = 0.1$, and $k = 36$.

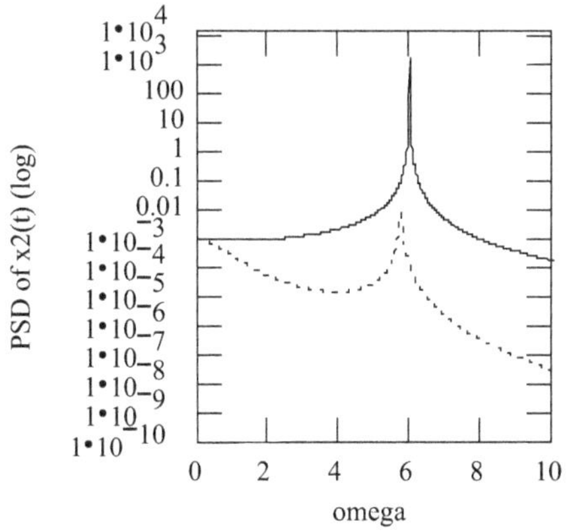


FIGURE 5.8 Plots of $S_{xx2}(\omega)$ (log) with $\beta = 0.1$ (solid), 1.9 (dot) when $m = 1$, $c = 0.1$, and $k = 36$ for $b = 0.2$.

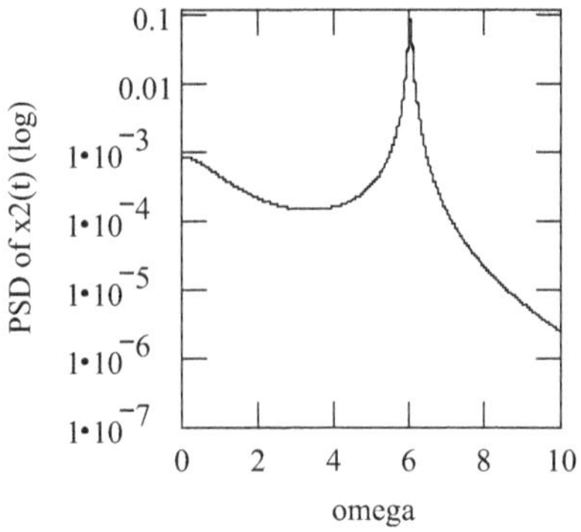


FIGURE 5.9 Plots of $S_{xx2}(\omega)$ (log) with $\beta = 1$, $m = 1$, $c = 0.1$, $k = 36$ for $b = 0.2$ (solid), 1 (dot), 10 (dash).

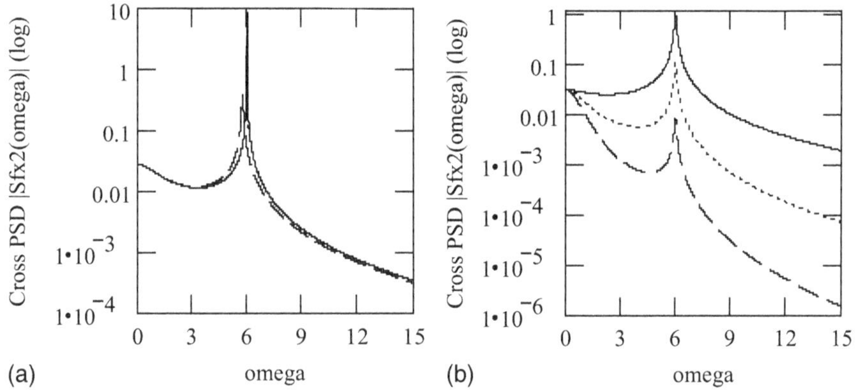


FIGURE 5.10 Plots of $|S_{fx2}(\omega)|$ when $m = 1$, $c = 0.1$, and $k = 36$. (a). $|S_{fx2}(\omega)|$ (log) with $(\beta, b) = (0.1, 0.5)$ (solid), $(1, 0.5)$ (dot), $(1.9, 0.5)$ (dash). (b). $|S_{fx2}(\omega)|$ (log) with $(\beta, b) = (1, 0.2)$ (solid), $(1, 0.8)$ (dot), $(1, 1.5)$ (dash).

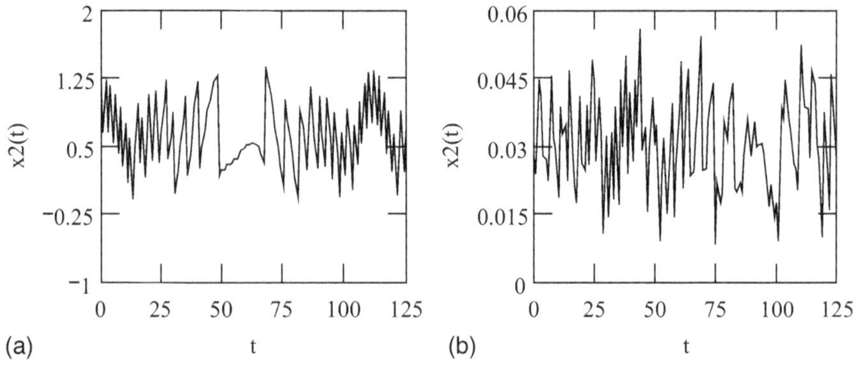


FIGURE 5.11 Response $x_2(t)$ when $m = 1$, $c = 0.1$, $k = 36$, and $b = 0.5$. (a). $x_2(t)$ for $\beta = 0.4$. (b). $x_2(t)$ for $\beta = 1.4$.

## 5.4 RESPONSES OF CLASS III FRACTIONAL VIBRATORS DRIVEN BY FRACTIONAL OU PROCESSES

### 5.4.1 Computations

The motion equation of a class III fractional vibrator is given by

$$m\frac{d^{\alpha}x_3(t)}{dt^{\alpha}}+c\frac{d^{\beta}x_3(t)}{dt^{\beta}}+kx_3(t)=f(t). \tag{5.16}$$

In (5.16), $x_3(t)$ is the response of a class III fractional vibrator.

**Theorem 5.5 (PSD response III)**

Let $S_{xx3}(\omega)$ be the PSD of $x_3(t)$. Then,

$$S_{xx3}(\omega)=\frac{\dfrac{1}{\left(\lambda^2+\omega^2\right)^b}}{k^2\left\{\left[1-\gamma^2\left(\omega^{\alpha-2}\left|\cos\dfrac{\alpha\pi}{2}\right|-2\varsigma\omega_n\omega^{\beta-2}\cos\dfrac{\beta\pi}{2}\right)\right]^2+\left[\dfrac{\gamma\left(\omega^{\alpha-1}\sin\dfrac{\alpha\pi}{2}+2\varsigma\omega_n\omega^{\beta-1}\sin\dfrac{\beta\pi}{2}\right)}{\omega_n\left(\omega^{\alpha-2}\left|\cos\dfrac{\alpha\pi}{2}\right|-2\varsigma\omega_n\omega^{\beta-2}\cos\dfrac{\beta\pi}{2}\right)}\right]^2\right\}}. \tag{5.17}$$

*Proof.* With $S_{xx3}(\omega) = S_{ff}(\omega)|H_3(\omega)|^2$, where $H_3(\omega)$ is expressed by (Chapter 7 in Volume I or Li [16, 17])

$$H_3(\omega)=\frac{1/k}{1-\gamma^2\left(\omega^{\alpha-2}\left|\cos\dfrac{\alpha\pi}{2}\right|-2\varsigma\omega_n\omega^{\beta-2}\cos\dfrac{\beta\pi}{2}\right)+i\dfrac{\gamma\left(\omega^{\alpha-1}\sin\dfrac{\alpha\pi}{2}+2\varsigma\omega_n\omega^{\beta-1}\sin\dfrac{\beta\pi}{2}\right)}{\omega_n\left(\omega^{\alpha-2}\left|\cos\dfrac{\alpha\pi}{2}\right|-2\varsigma\omega_n\omega^{\beta-2}\cos\dfrac{\beta\pi}{2}\right)}}, \tag{5.18}$$

we have (5.17). The proof is finished.

**Theorem 5.6 (cross-PSD response III)**

Let $S_{fx3}(\omega)$ be the cross PSD between $f(t)$ and $x_3(t)$. Then,

$$S_{fx3}(\omega) = \frac{\dfrac{1}{\left(\lambda^2+\omega^2\right)^b}}{1-\gamma^2\left(\omega^{\alpha-2}\left|\cos\dfrac{\alpha\pi}{2}\right|-2\varsigma\omega_n\omega^{\beta-2}\cos\dfrac{\beta\pi}{2}\right)+i\dfrac{\gamma\left(\omega^{\alpha-1}\sin\dfrac{\alpha\pi}{2}+2\varsigma\omega_n\omega^{\beta-1}\sin\dfrac{\beta\pi}{2}\right)}{\omega_n\left(\omega^{\alpha-2}\left|\cos\dfrac{\alpha\pi}{2}\right|-2\varsigma\omega_n\omega^{\beta-2}\cos\dfrac{\beta\pi}{2}\right)}}. \tag{5.19}$$

*Proof.* Using $S_{fx3}(\omega) = S_{ff}(\omega)H_3(\omega)$ and (5.18) yields (5.19). The proof ends.

Let $r_{xx3}(\tau)$ be the ACF of $x_3(t)$. Let $r_{fx3}(\tau)$ be the cross-correlation between the excitation $f(t)$ and the response $x_3(t)$. Denote by $h_3(\tau)$ the impulse response of a class III fractional vibrator. Applying the convolution theory to $S_{xx3}(\omega) = S_{ff}(\omega)|H_3(\omega)|^2$ produces the ACF response $r_{xx3}(\tau)$ expressed by

$$r_{xx3}(\tau) = r_{ff}(\tau)*h_3(\tau)*h_3(-\tau). \tag{5.20}$$

Similarly, applying the convolution theory to $S_{fx3}(\omega) = S_{ff}(\omega)H_3(\omega)$ results in the cross-correlation response

$$r_{fx3}(\tau) = r_{ff}(\tau)*h_3(\tau). \tag{5.21}$$

In (5.21), where $h_3(\tau)$ is given by (Chapter 7 in Volume I or Li [16, 17])

$$h_3(t) = \frac{e^{-\frac{m\omega^{\alpha-1}\sin\frac{\alpha\pi}{2}+c\omega^{\beta-1}\sin\frac{\beta\pi}{2}}{2\sqrt{-\left(m\omega^{\alpha-2}\cos\frac{\alpha\pi}{2}+c\omega^{\beta-2}\cos\frac{\beta\pi}{2}\right)k}}\omega_{\text{eqn}3}t}\sin\omega_{\text{eqd}3}t}{-\left(m\omega^{\alpha-2}\cos\dfrac{\alpha\pi}{2}+c\omega^{\beta-2}\cos\dfrac{\beta\pi}{2}\right)\omega_{\text{eqd}3}}u(t). \tag{5.22}$$

Refer to Chapter 7 in Volume I or Li [16, 17] for the expressions of $\omega_{\text{eqn}3}$ and $\omega_{\text{eqd}3}$ in (5.22).

### 5.4.2 Effect of $(\alpha, \beta)$ on Responses

Some plots of $|H_3(\omega)|^2$ are shown in Figure 5.5. Figure 5.13 indicates some plots of $S_{xx3}(\omega)$. Some plots of $|S_{fx3}(\omega)|$ are given in Figure 5.14. Figures 5.13 and 5.14 exhibit that the effect of $(\alpha, \beta)$ on the responses to class III fractional vibration systems driven by the fractional OU processes is significant.

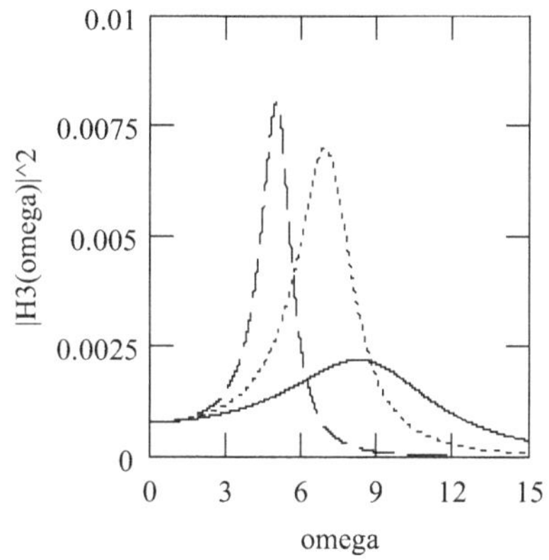


FIGURE 5.12 Plots of $|H_3(\omega)|^2$ when $m = 1, c = 0.1$, and $k = 36$, for $(\alpha, \beta) = (1.6, 0.1)$ (solid), (1.8, 1.8) (dot), (2.2, 0.1) (dash).

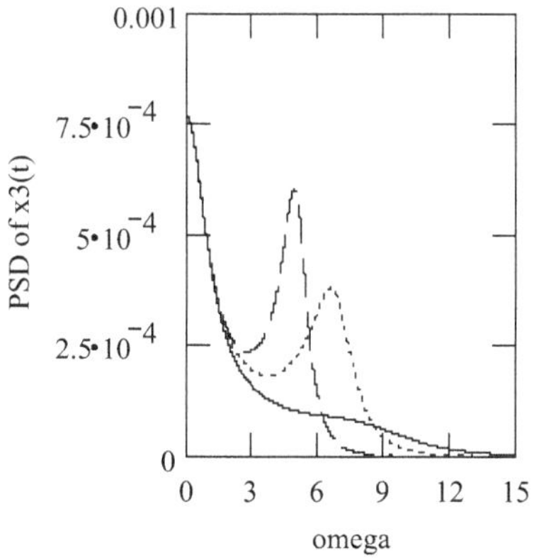


FIGURE 5.13 Plots of $S_{xx3}(\omega)$ when $m = 1, c = 0.1$, and $k = 36$ when $(\alpha, \beta) = (1.6, 0.1)$ (solid), (1.8, 1.8) (dot), (2.2, 0.1) (dash) for $b = 0.8$.

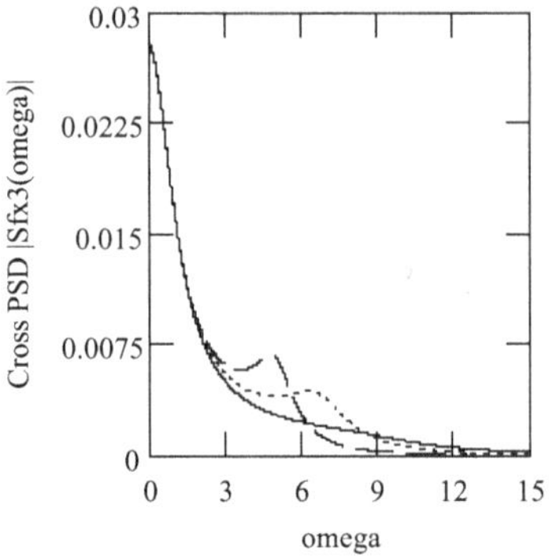


FIGURE 5.14 Plots of $|S_{fx3}(\omega)|$ when $m=1, c=0.1$, and $k=36$ for $(\alpha,\beta)=(1.6,0.1)$ (solid), (1.8, 1.8) (dot), (2.2, 0.1) (dash) when $b=0.8$.

## 5.5 RESPONSES OF CLASS IV FRACTIONAL VIBRATION SYSTEMS DRIVEN BY FRACTIONAL OU PROCESSES

### 5.5.1 Computations

The motion equation of a class IV fractional vibrator is given by

$$m\frac{d^\alpha x_4(t)}{dt^\alpha}+k\frac{d^\lambda x_4(t)}{dt^\lambda}=f(t). \tag{5.23}$$

In (4.23), $0\le\lambda<1$ and $x_4(t)$ is the response of a class IV fractional vibrator.

**Theorem 5.7 (PSD response IV)**

Let $S_{xx4}(\omega)$ be the PSD of $x_4(t)$. Then,

$$S_{xx4}(\omega)=\frac{\dfrac{1}{\left(\lambda^2+\omega^2\right)^b}}{k^2\omega^{2\lambda}\cos^2\dfrac{\lambda\pi}{2}\left[\left(1-\gamma^2\dfrac{-\omega^{\alpha-2}\cos\dfrac{\alpha\pi}{2}}{\omega^\lambda\cos\dfrac{\lambda\pi}{2}}\right)^2+4\left(\gamma\dfrac{m\omega^{\alpha-1}\sin\dfrac{\alpha\pi}{2}+k\omega^{\lambda-1}\sin\dfrac{\lambda\pi}{2}}{2\sqrt{mk\omega^{\alpha+\lambda-2}\left|\cos\dfrac{\alpha\pi}{2}\right|\cos\dfrac{\lambda\pi}{2}}}\sqrt{\dfrac{-\omega^{\alpha-2}\cos\dfrac{\alpha\pi}{2}}{\omega^\lambda\cos\dfrac{\lambda\pi}{2}}}\right)^2\right]}. \tag{5.24}$$

*Proof.* Performing the operation of $S_{xx4}(\omega) = S_{ff}(\omega)|H_4(\omega)|^2$, where $H_4(\omega)$ is given by (Chapter 7 in Volume I or Li [16, 17])

$$H_4(\omega)=\frac{1}{k\omega^{\lambda}\cos\frac{\lambda\pi}{2}\left(1-\gamma^2\frac{-\omega^{\alpha-2}\cos\frac{\alpha\pi}{2}}{\omega^{\lambda}\cos\frac{\lambda\pi}{2}}+i2\gamma\frac{m\omega^{\alpha-1}\sin\frac{\alpha\pi}{2}+k\omega^{\lambda-1}\sin\frac{\lambda\pi}{2}}{2\sqrt{mk\omega^{\alpha+\lambda-2}\left|\cos\frac{\alpha\pi}{2}\right|\cos\frac{\lambda\pi}{2}}}\sqrt{\frac{-\omega^{\alpha-2}\cos\frac{\alpha\pi}{2}}{\omega^{\lambda}\cos\frac{\lambda\pi}{2}}}\right)}, \tag{5.25}$$

we have (5.24). This finishes the proof.

**Theorem 5.8 (cross-PSD response IV)**

Denote by $S_{fx4}(\omega)$ the cross-PSD between $f(t)$ and $x_4(t)$. Then,

$$S_{fx4}(\omega)=\frac{\frac{1}{\left(\lambda^2+\omega^2\right)^b}}{k\omega^{\lambda}\cos\frac{\lambda\pi}{2}\left(1-\gamma^2\frac{-\omega^{\alpha-2}\cos\frac{\alpha\pi}{2}}{\omega^{\lambda}\cos\frac{\lambda\pi}{2}}+i2\gamma\frac{m\omega^{\alpha-1}\sin\frac{\alpha\pi}{2}+k\omega^{\lambda-1}\sin\frac{\lambda\pi}{2}}{2\sqrt{mk\omega^{\alpha+\lambda-2}\left|\cos\frac{\alpha\pi}{2}\right|\cos\frac{\lambda\pi}{2}}}\sqrt{\frac{-\omega^{\alpha-2}\cos\frac{\alpha\pi}{2}}{\omega^{\lambda}\cos\frac{\lambda\pi}{2}}}\right)}. \tag{5.26}$$

*Proof.* Doing the operation of $S_{fx4}(\omega) = S_{ff}(\omega)H_4(\omega)$ and (5.25) produces (5.26). This finishes the proof.

Let $h_4(\tau)$ be the impulse response of a class IV fractional vibrator. Doing the inverse Fourier transform on both sides of $S_{xx4}(\omega) = S_{ff}(\omega)|H_4(\omega)|^2$ yields

$$r_{xx4}(\tau) = r_{ff}(\tau) * h_4(\tau) * h_4(-\tau), \tag{5.27}$$

where $r_{xx4}(\tau)$ is the ACF of $x_4(t)$. In addition, performing the inverse Fourier transform on both sides of $S_{fx4}(\omega) = S_{ff}(\omega)H_4(\omega)$ produces

$$r_{fx4}(\tau) = r_{ff}(\tau) * h_4(\tau), \tag{5.28}$$

where $r_{fx4}(\tau)$ is the cross-correlation between $f(t)$ and $x_4(t)$. In (5.27) and (5.28) (Chapter 7 in Volume I or Li [16, 17]),

$$h_4(t) = e^{-\frac{m\omega^{\alpha-1}\sin\frac{\alpha\pi}{2}+k\omega^{\lambda-1}\sin\frac{\lambda\pi}{2}}{2\sqrt{mk\omega^{\alpha+\lambda-2}\left|\cos\frac{\alpha\pi}{2}\right|\cos\frac{\lambda\pi}{2}}}\sqrt{\frac{\omega^{\lambda}\cos\frac{\lambda\pi}{2}}{-\omega^{\alpha-2}\cos\frac{\alpha\pi}{2}}}\omega_n t} \frac{1}{m_{\mathrm{eq4}}\omega_{\mathrm{eqd4}}}\sin\omega_{\mathrm{eqd4}} t u(t). \tag{5.29}$$

Refer to Chapter 7 in Volume I or Li [16, 17] for the expressions of $m_{\mathrm{eq4}}$, $\omega_{\mathrm{eqn4}}$, and $\omega_{\mathrm{eqd4}}$ in (5.29).

## 5.5.2 Effect of $(\alpha, \lambda)$ on Responses

Figure 5.15 illustrates some plots of $|H_4(\omega)|^2$. Figure 5.16 indicates some plots of $S_{xx4}(\omega)$. Some plots of $S_{fx4}(\omega)$ are shown in Figure 5.17. Figures 5.16 and 5.17

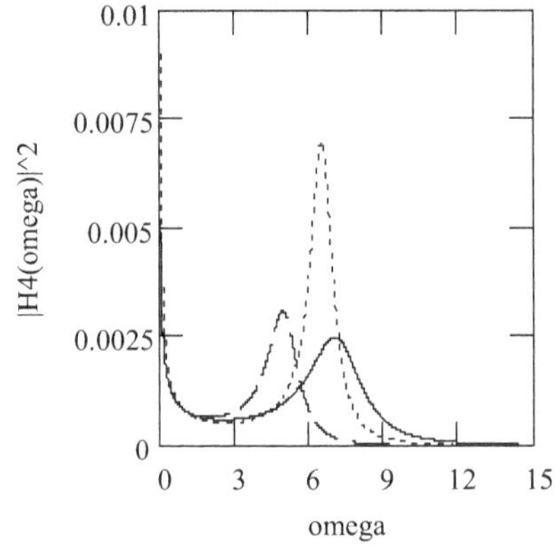


FIGURE 5.15 Plots of $|H_4(\omega)|^2$ when $m = 1$, $c = 0$, and $k = 36$, for $(\alpha, \lambda) = (2.1, 0.2)$ (solid), (2.3, 0.4) (dot), (2.3, 0.3) (dash).

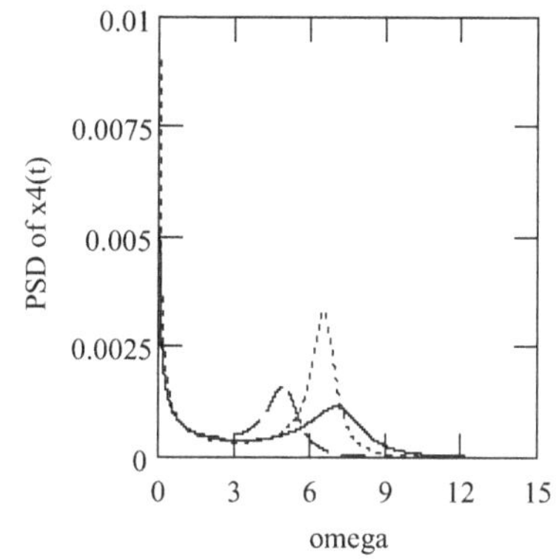


FIGURE 5.16 Plots of $S_{xx4}(\omega)$ when $m = 1$, $c = 0$, and $k = 36$ for $(\alpha, \lambda) = (2.1, 0.2)$ (solid), (2.3, 0.4) (dot), (2.5, 0.3) (dash) for $b = 0.2$.

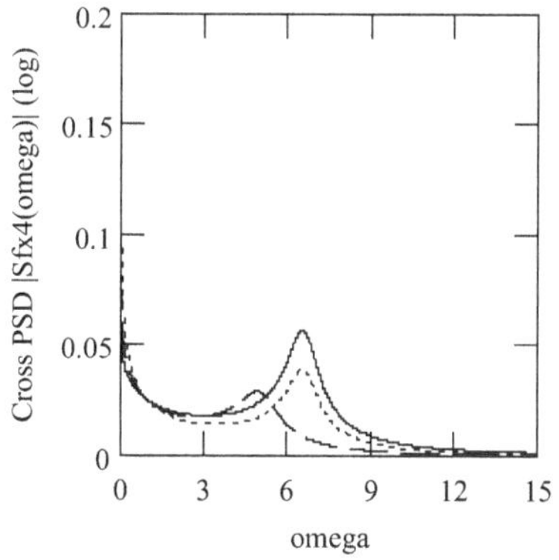


FIGURE 5.17 Plots of $|S_{fx4}(\omega)|$ when $m = 1$, $c = 0$, $k = 36$ for $(\alpha, \lambda) = (2.1, 0.2)$ (solid), (2.3, 0.4) (dot), (2.5, 0.3) (dash) for $b = 0.2$.

exhibit that the effects of $(\alpha, \lambda)$ and $b$ on the responses to class IV fractional vibration systems driven by the fractional OU processes are noticeable.

## 5.6 RESPONSES OF CLASS V FRACTIONAL VIBRATORS DRIVEN BY FRACTIONAL OU PROCESSES

### 5.6.1 Computation Methods

The motion equation of a class V fractional vibrator is in the form

$$m\frac{d^2x_5(t)}{dt^2}+k\frac{d^\lambda x_5(t)}{dt^\lambda}=f(t). \tag{5.30}$$

In (5.30), $x_5(t)$ is the response of a class V fractional vibrator.

**Theorem 5.9 (PSD response V)**

Denote by $S_{xx5}(\omega)$ the PSD of $x_5(t)$. Then,

$$S_{xx5}(\omega)=\frac{\frac{1}{\left(\lambda^2+\omega^2\right)^b}}{k^2\omega^{2\lambda}\cos^2\frac{\lambda\pi}{2}\left[\left(1-\frac{\gamma^2}{\omega^\lambda\cos\frac{\lambda\pi}{2}}\right)^2+4\gamma^2\left(\frac{k\omega^{\lambda-1}\sin\frac{\lambda\pi}{2}}{2\sqrt{mk\omega^\lambda\cos\frac{\lambda\pi}{2}}}\sqrt{\frac{1}{\omega^\lambda\cos\frac{\lambda\pi}{2}}}\right)^2\right]}. \tag{5.31}$$

*Proof.* Doing the operation of $S_{xx5}(\omega) = S_{ff}(\omega)|H_5(\omega)|^2$ and considering $H_5(\omega)$ in the form of (5.32) (Chapter 7 in Volume I or Li [16, 17])

$$H_5(\omega)=\frac{1}{k\omega^\lambda\cos\frac{\lambda\pi}{2}\left(1-\frac{\gamma^2}{\omega^\lambda\cos\frac{\lambda\pi}{2}}+i2\gamma\frac{k\omega^{\lambda-1}\sin\frac{\lambda\pi}{2}}{2\sqrt{mk\omega^\lambda\cos\frac{\lambda\pi}{2}}}\sqrt{\frac{1}{\omega^\lambda\cos\frac{\lambda\pi}{2}}}\right)}, \tag{5.32}$$

we have (5.31). The proof is finished.

**Theorem 5.10 (cross-PSD response V)**

Let $S_{fx5}(\omega)$ be the cross-PSD between $f(t)$ and $x_5(t)$. Then,

$$S_{fx5}(\omega)=\frac{\frac{1}{\left(\lambda^2+\omega^2\right)^b}}{k\omega^\lambda\cos\frac{\lambda\pi}{2}\left(1-\frac{\gamma^2}{\omega^\lambda\cos\frac{\lambda\pi}{2}}+i2\gamma\frac{k\omega^{\lambda-1}\sin\frac{\lambda\pi}{2}}{2\sqrt{mk\omega^\lambda\cos\frac{\lambda\pi}{2}}}\sqrt{\frac{1}{\omega^\lambda\cos\frac{\lambda\pi}{2}}}\right)}. \tag{5.33}$$

*Proof.* Doing the operation of $S_{fx5}(\omega) = S_{ff}(\omega)H_5(\omega)$ with the consideration of (5.32) yields (5.33) (Li [18]). This finishes the proof.

Let $h_5(t)$ be the impulse response of a class V fractional vibrator. Performing the inverse Fourier transform on both sides of $S_{xx5}(\omega) = S_{ff}(\omega)|H_5(\omega)|^2$ produces

$$r_{xx5}(\tau) = r_{ff}(\tau) * h_5(\tau) * h_5(-\tau), \tag{5.34}$$

where $r_{xx5}(\tau)$ is the ACF of $x_5(t)$. In addition, doing the inverse Fourier transform on both sides of $S_{fx5}(\omega) = S_{ff}(\omega)H_5(\omega)$ yields

$$r_{fx5}(\tau) = r_{ff}(\tau) * h_5(\tau), \tag{5.35}$$

where $r_{fx5}(\tau)$ is the cross-correlation between $f(t)$ and $x_5(t)$. In (5.34) and (5.35) (Chapter 7 in Volume I or Li [16, 17]),

$$h_5(t) = e^{-\frac{k\omega^{\lambda-1}\sin\frac{\lambda\pi}{2}}{2\sqrt{mk\omega^{\lambda}\cos\frac{\lambda\pi}{2}}}\sqrt{\omega^{\lambda}\cos\frac{\lambda\pi}{2}}\,\omega_n t} \frac{1}{m\omega_{\text{eqd5}}} \sin \omega_{\text{eqd5}} t u(t). \tag{5.36}$$

Refer to Chapter 7 in Volume I or Li [16, 17] for the expression of $\omega_{\text{eqd5}}$ in (5.36).

### 5.6.2 Effect of $\lambda$ on Responses

Figure 5.18 illustrates some plots of $|H_5(\omega)|^2$. Figure 5.19 indicates some plots of $S_{xx5}(\omega)$. Figure 5.20 shows some plots of $|S_{fx5}(\omega)|$. From Figures 5.19

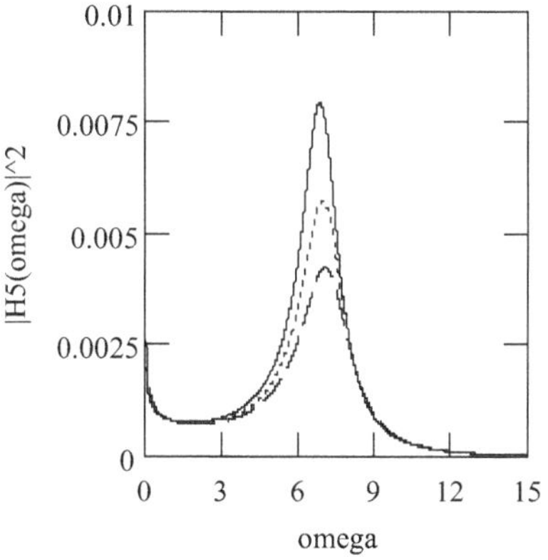


FIGURE 5.18 Plots of $|H_5(\omega)|^2$ when $m = 1$, $c = 0$, and $k = 36$, for $\lambda = 0.15$ (solid), 0.17 (dot), 0.19 (dash).

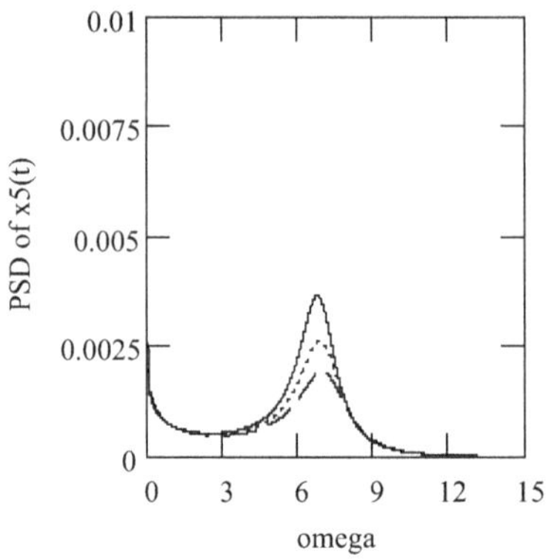


FIGURE 5.19 Plots of $S_{xx5}(\omega)$ when $m = 1$, $c = 0$, and $k = 36$ for $\lambda = 0.15$ (solid), 0.17 (dot), 0.19 (dash) for $b = 0.2$.

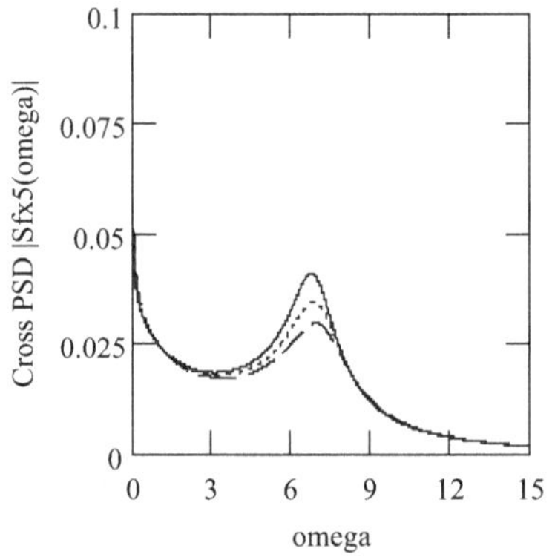


FIGURE 5.20 Plots of $|S_{fx5}(\omega)|$ when $m = 1$, $c = 0$, and $k = 36$ for $\lambda = 0.15$ (solid), 0.17 (dot), 0.19 (dash) when $b = 0.2$.

and 5.20, we see that the effects of $\lambda$ and $b$ on the responses to class V fractional vibrators driven by the fractional OU processes are considerable.

## 5.7 RESPONSES OF CLASS VI FRACTIONAL VIBRATORS DRIVEN BY FRACTIONAL OU PROCESSES

### 5.7.1 Computations

Consider the motion equation of a class VI fractional vibrator

$$m\frac{d^{\alpha}x_6(t)}{dt^{\alpha}}+c\frac{d^{\beta}x_6(t)}{dt^{\beta}}+k\frac{d^{\lambda}x_6(t)}{dt^{\lambda}}=f(t). \tag{5.37}$$

In (5.37), $x_6(t)$ is the response of a class VI fractional vibrator.

**Theorem 5.11 (PSD response VI)**

Let $S_{xx6}(\omega)$ be the PSD of $x_6(t)$. Then,

$$S_{xx6}(\omega)=\frac{1}{k^2}\frac{\dfrac{1}{\left(\lambda^2+\omega^2\right)^b}}{\left[\omega^\lambda\cos\dfrac{\lambda\pi}{2}+\gamma^2\left(\omega^{\alpha-2}\cos\dfrac{\alpha\pi}{2}+2\varsigma\omega_n\omega^{\beta-2}\cos\dfrac{\beta\pi}{2}\right)\right]^2+\gamma^2\left(\omega^{\alpha-1}\sin\dfrac{\alpha\pi}{2}+2\zeta\omega_n\omega^{\beta-1}\sin\dfrac{\beta\pi}{2}+\omega_n^2\omega^{\lambda-1}\sin\dfrac{\lambda\pi}{2}\right)^2}. \tag{5.38}$$

*Proof.* Following Chapter 7 in Volume I or Li [16, 17], $H_6(\omega)$ is given by

$$H_6(\omega)=\frac{1}{k\left[\begin{array}{l}\omega^\lambda\cos\dfrac{\lambda\pi}{2}+\gamma^2\left(\omega^{\alpha-2}\cos\dfrac{\alpha\pi}{2}+2\varsigma\omega_n\omega^{\beta-2}\cos\dfrac{\beta\pi}{2}\right)\\+i\gamma\left(\omega^{\alpha-1}\sin\dfrac{\alpha\pi}{2}+2\zeta\omega_n\omega^{\beta-1}\sin\dfrac{\beta\pi}{2}+\omega_n^2\omega^{\lambda-1}\sin\dfrac{\lambda\pi}{2}\right)\end{array}\right]}. \tag{5.39}$$

Thus, doing the operation of $S_{xx6}(\omega)=S_{ff}(\omega)|H_6(\omega)|^2$ yields (5.38). The proof ends.

**Theorem 5.12 (cross-PSD response VI)**

Denote by $S_{fx6}(\omega)$ the cross-PSD between $f(t)$ and $x_6(t)$. Then,

$$S_{fx6}(\omega)=\frac{\dfrac{1}{\left(\lambda^2+\omega^2\right)^b}}{k\left[\begin{array}{l}\omega^\lambda\cos\dfrac{\lambda\pi}{2}+\gamma^2\left(\omega^{\alpha-2}\cos\dfrac{\alpha\pi}{2}+2\varsigma\omega_n\omega^{\beta-2}\cos\dfrac{\beta\pi}{2}\right)\\+i\gamma\left(\omega^{\alpha-1}\sin\dfrac{\alpha\pi}{2}+2\zeta\omega_n\omega^{\beta-1}\sin\dfrac{\beta\pi}{2}+\omega_n^2\omega^{\lambda-1}\sin\dfrac{\lambda\pi}{2}\right)\end{array}\right]}. \tag{5.40}$$

*Proof.* Performing the operation of $S_{fx6}(\omega)=S_{ff}(\omega)H_6(\omega)$ and taking into account (5.39) results in (5.40). The proof is finished.

Let $h_6(\tau)$ be the impulse response of a class VI fractional vibrator. Doing the inverse Fourier transform on both sides of $S_{xx6}(\omega)=S_{ff}(\omega)|H_6(\omega)|^2$ produces

$$r_{xx6}(\tau) = r_{ff}(\tau) * h_6(\tau) * h_6(-\tau), \tag{5.41}$$

where $r_{xx6}(\tau)$ is the ACF of $x_6(t)$. Besides,

$$r_{fx6}(\tau) = r_{ff}(\tau) * h_6(\tau), \tag{5.42}$$

where $r_{fx6}(\tau)$ is the cross-correlation between $f(t)$ and $x_6(t)$. In (5.41) and (5.42) (Chapter 7 in Volume I or Li [16, 17]),

$$h_6(t) = e^{-\zeta_{eq6}\omega_{eqn6}t} \frac{1}{m_{eq6}\omega_{eqd6}} \sin \omega_{eqd6} t u(t). \tag{5.43}$$

Refer to Chapter 7 in Volume I or Li [16, 17] for the expressions of $\zeta_{eq6}$, $m_{eq6}$, $\omega_{eqn6}$, and $\omega_{eqd6}$ in (5.43).

### 5.7.2 Effect of $(\alpha, \beta, \lambda)$ on Responses

Figure 5.21 illustrates some plots of $|H_6(\omega)|^2$. Figure 5.22 shows some plots of $S_{xx6}(\omega)$. Figure 5.23 demonstrates some plots of $|S_{fx6}(\omega)|$. Figures 5.21 and 5.22 show that the effect of $(\alpha, \beta, \lambda)$ on the responses to class VI fractional vibration systems driven by the fractional OU processes is significant.

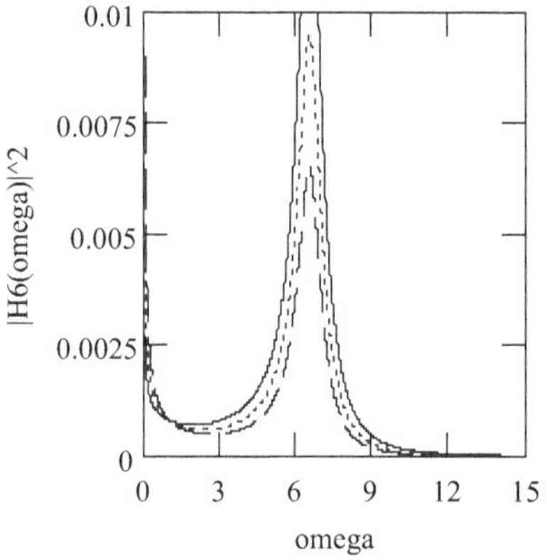


FIGURE 5.21 Plots of $|H_6(\omega)|^2$ when $m = 1$, $c = 0.1$, and $k = 36$, for $(\alpha, \beta, \lambda) =$ (2.1, 0.7, 0.2) (solid), (2.2, 0.8, 0.3) (dot), (2.3, 0.9, 0.4) (dash).

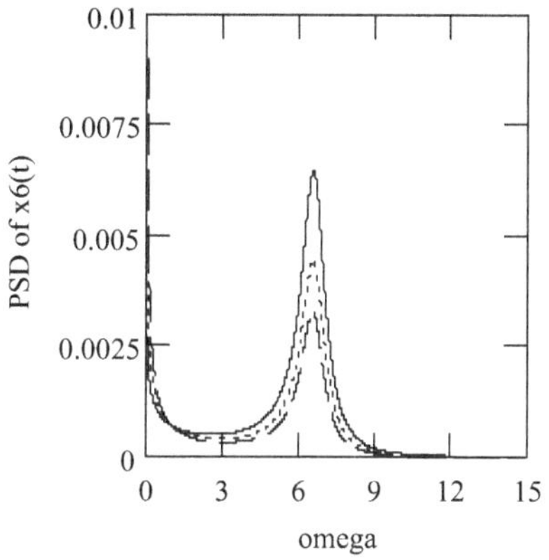


FIGURE 5.22 Plots of PSD response $S_{xx6}(\omega)$ when $m = 1$, $c = 0.1$, $k = 36$ for $(\alpha, \beta, \lambda) = (2.1, 0.7, 0.2)$ (solid), $(2.2, 0.8, 0.3)$ (dot), $(2.3, 0.9, 0.4)$ (dash) when $b = 0.2$.

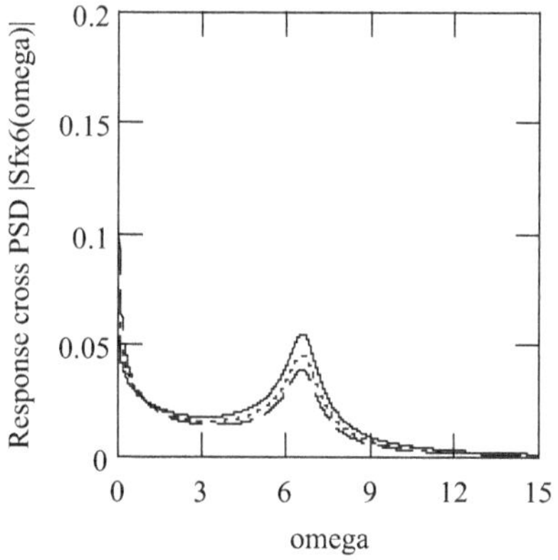


FIGURE 5.23 Plots of $|S_{fx6}(\omega)|$ when $m = 1, c = 0.1, k = 36$ for $(\alpha, \beta, \lambda) = (2.1, 0.7, 0.2)$ (solid), $(2.2, 0.8, 0.3)$ (dot), $(2.3, 0.9, 0.4)$ (dash) when $b = 0.2$.

## 5.8 RESPONSES OF CLASS VII FRACTIONAL VIBRATORS DRIVEN BY FRACTIONAL OU PROCESSES

### 5.8.1 Computations

Consider the motion equation of a class VII fractional vibrator

$$m\frac{d^2 x_7(t)}{dt^2} + c\frac{d^\beta x_7(t)}{dt^\beta} + k\frac{d^\lambda x_7(t)}{dt^\lambda} = f(t). \tag{5.44}$$

In (5.44), $x_7(t)$ is the response of a class VII fractional vibrator.

**Theorem 5.13 (PSD response VII)**

Denote by $S_{xx7}(\omega)$ the PSD of $x_7(t)$. Then,

$$S_{xx7}(\omega)=\frac{1}{k^2}\frac{\dfrac{1}{\left(\lambda^2+\omega^2\right)^b}}{\left[\omega^\lambda\cos\dfrac{\lambda\pi}{2}-\gamma\left(1-2\varsigma\omega_n\omega^{\beta-2}\cos\dfrac{\beta\pi}{2}\right)\right]^2+\gamma^2\left(2\varsigma\omega^{\beta-1}\sin\dfrac{\beta\pi}{2}+\omega_n\omega^{\lambda-1}\sin\dfrac{\lambda\pi}{2}\right)^2}. \tag{5.45}$$

*Proof.* Doing the operation of $S_{xx7}(\omega)=S_{ff}(\omega)|H_7(\omega)|^2$ and considering $H_7(\omega)$ given by (Chapter 7 in Volume I or Li [17])

$$H_7(\omega)=\frac{1}{k\left[\begin{aligned}&\omega^\lambda\cos\frac{\lambda\pi}{2}-\gamma\left(1-2\varsigma\omega_n\omega^{\beta-2}\cos\frac{\beta\pi}{2}\right)\\&+i\gamma\left(2\varsigma\omega^{\beta-1}\sin\frac{\beta\pi}{2}+\omega_n\omega^{\lambda-1}\sin\frac{\lambda\pi}{2}\right)\end{aligned}\right]}, \tag{5.46}$$

results in (5.45). The proof completes.

**Theorem 5.14 (cross-PSD response VII)**

Let $S_{fx7}(\omega)$ be the cross PSD between $f(t)$ and $x_7(t)$. Then,

$$S_{fx7}(\omega)=\frac{\dfrac{1}{\left(\lambda^2+\omega^2\right)^b}}{k\left[\begin{aligned}&\omega^\lambda\cos\frac{\lambda\pi}{2}-\gamma\left(1-2\varsigma\omega_n\omega^{\beta-2}\cos\frac{\beta\pi}{2}\right)\\&+i\gamma\left(2\varsigma\omega^{\beta-1}\sin\frac{\beta\pi}{2}+\omega_n\omega^{\lambda-1}\sin\frac{\lambda\pi}{2}\right)\end{aligned}\right]}. \tag{5.47}$$

*Proof.* Doing the operation of $S_{fx7}(\omega)=S_{ff}(\omega)H_7(\omega)$ with the consideration of (5.46) yields (5.47). The proof ends.

Denote by $h_7(t)$ the impulse response of a class VII fractional vibrator. Doing the inverse Fourier transform on both sides of $S_{xx7}(\omega) = S_{ff}(\omega)|H_7(\omega)|^2$ produces

$$r_{xx7}(\tau) = r_{ff}(\tau) * h_7(\tau) * h_7(-\tau), \tag{5.48}$$

where $r_{xx7}(\tau)$ is the ACF of $x_7(t)$. Additionally,

$$r_{fx7}(\tau) = r_{ff}(\tau) * h_7(\tau), \tag{5.49}$$

where $r_{fx7}(\tau)$ is the cross-correlation between $f(t)$ and $x_7(t)$. In (5.48) and (5.49) (Chapter 7 in Volume I or Li [17]),

$$h_7(t) = e^{-\zeta_{eq7}\omega_{eqn7}t} \frac{1}{m_{eq7}\omega_{eqd7}} \sin \omega_{eqd7} t, \quad t \geq 0. \tag{5.50}$$

Refer to Chapter 7 in Volume I or Li [17] for the expressions of $\zeta_{eq7}$, $m_{eq7}$, $\omega_{eqn7}$, and $\omega_{eqd7}$ in (5.50).

### 5.8.2 Effect of ($\beta$, $\lambda$) on Responses

Figure 5.24 shows some plots of $|H_7(\omega)|^2$. Figure 5.25 indicates the plots of $S_{xx7}(\omega)$. Figure 5.26 illustrates some plots of $|S_{fx7}(\omega)|$. Figures 5.25 and

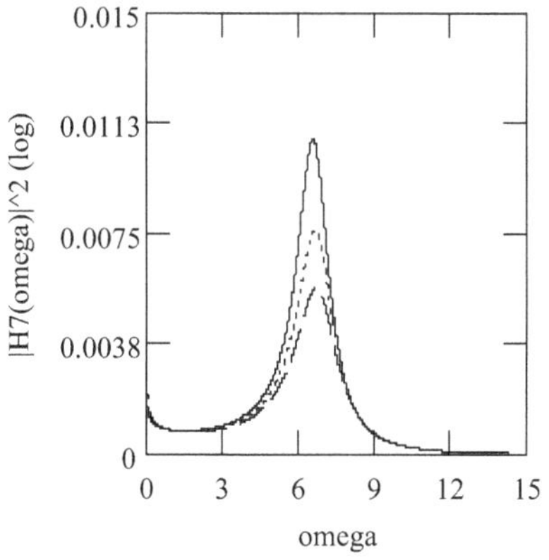


FIGURE 5.24 Plots of $|H_7(\omega)|^2$ when $m = 1$, $c = 0.1$, and $k = 36$, for $(\beta, \lambda) = (1.5, 0.12)$ (solid), (1.6, 0.14) (dot), (1.6, 0.16) (dash).

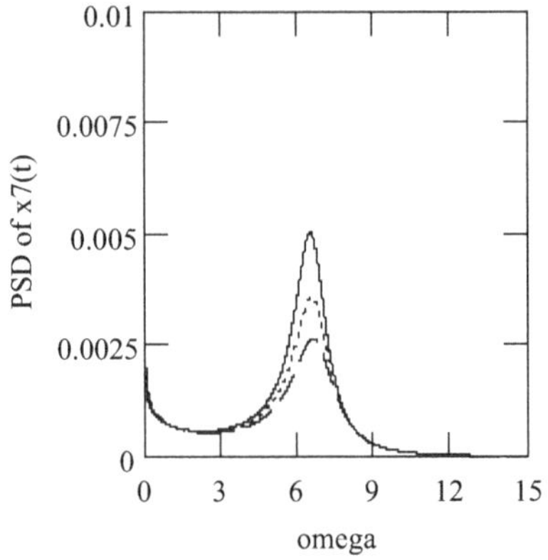


FIGURE 5.25 Plots of response PSD $S_{xx7}(\omega)$ when $m = 1$, $c = 0.1$, and $k = 36$ for $(\beta, \lambda) = (1.5, 0.12)$ (solid), (1.6, 0.14) (dot), (1.6, 0.16) (dash) for $b = 0.2$.

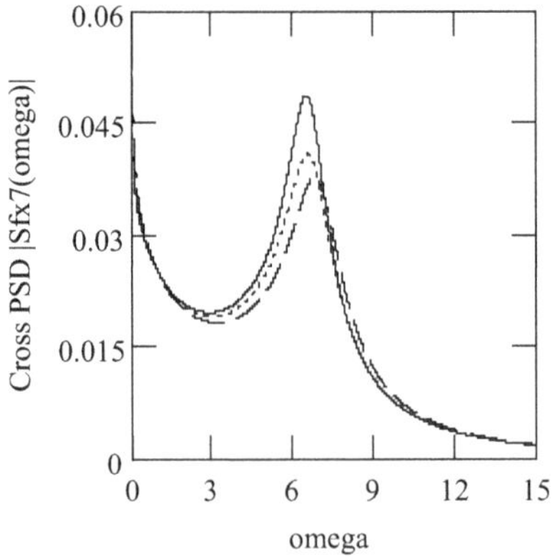


FIGURE 5.26 Plots of cross-PSD response $|S_{fx7}(\omega)|$ when $m = 1$, $c = 0.1$, and $k = 36$ for $(\beta, \lambda) = (1.5, 0.12)$ (solid), (1.6, 0.14) (dot), (1.6, 0.16) (dash) for $b = 0.2$.

5.26 exhibit that the effect of $(\beta, \lambda)$ on the responses to class VII fractional vibration systems driven by the fractional class VII vibrators is considerable.

## 5.9 SUMMARY

The analytic expressions of the PSD and cross-PSD responses to seven classes of fractional vibrators driven by the fractional OU processes have been presented in Theorems 5.1–5.14, respectively. We have illustrated that there are effects of orders of fractional vibrators on the responses considerably. The responses of seven classes of fractional vibrators under the excitation of the fractional OU processes are short-range dependent due to $S_{xxi}(0) < \infty$ for $i = 1, \ldots, 7$.

## 5.10 EXERCISES

5.1. Let $S_{ff}(\omega) = \dfrac{1}{\left(\lambda^2 + \omega^2\right)^b}$, where $b > 0$, $\lambda > 0$. Find the inverse Fourier transform of $S_{ff}(\omega)$.

5.2. Let $r_{ff}(\tau)$ be the inverse Fourier transform of $S_{ff}(\omega)$ in Exercise 5.1. Find the condition for $\int_{-\infty}^{\infty} r_{ff}(\tau) = \infty$.

5.3. Find the condition for $\int_{-\infty}^{\infty} r_{ff}(\tau) < \infty$, where $r_{ff}(\tau) = F^{-1}[S_{ff}(\omega)]$ in Exercise 5.1.

5.4. Following Chapter 7 in Volume I, we write

$$h_7(t) = e^{-\varsigma_{eq7}\omega_{eqn7}t} \frac{1}{m_{eq7}\omega_{eqd7}} \sin \omega_{eqd7} t, \quad t \geq 0.$$

Find $h_7(\tau) * h_7(-\tau)$.

5.5. Find $r_{xx7}(\tau) = r_{ff}(\tau) * h_7(\tau) * h_7(-\tau)$, where $r_{ff}(\tau) = F^{-1}[S_{ff}(\omega)]$ in Exercise 5.1.

5.6. Find $r_{fx7}(\tau) = r_{ff}(\tau) * h_7(\tau)$, where $r_{ff}(\tau) = F^{-1}[S_{ff}(\omega)]$ in Exercise 5.1.

5.7. Find the Fourier transform of $r_{fx7}(\tau)$.

5.8. Find the condition for $\int_{-\infty}^{\infty} r_{xx7}(\tau) < \infty$.

5.9. Find the condition for $\int_{-\infty}^{\infty} r_{xx7}(\tau) = \infty$.

## REFERENCES

1. G. A. Pavliotis, *Stochastic Processes and Applications, Diffusion Processes, the Fokker-Planck and Langevin Equations, Texts in Applied Mathematics*, Vol. 60, Springer, New York, 2014.

2. W. T. Coffey, Yu P. Kalmykov, and J. T. Waldron, *The Langevin Equation with Applications to Stochastic Problems in Physics, Chemistry and Electrical Engineering*, 2nd ed., World Scientific, Singapore, 2004.
3. I. I. Eliazar and M. F. Shlesinger, Fractional motions, *Physics Reports*, 527(2): 2013, 101–129.
4. B. J. West, E. L. Geneston, and P. Grigolini, Maximizing information exchange between complex networks, *Physics Reports*, 468(1–3): 2008, 1–99.
5. C. H. Eab and S. C. Lim, Fractional generalized Langevin equation approach to single-file diffusion, *Physica A*, 389(13): 2010, 2510–2521.
6. C. H. Eab and S. C. Lim, Fractional Langevin equations of distributed order, *Physcial Review E*, 83(3): 2011, 031136, 10.
7. S. C. Lim, M. Li, and L. P. Teo, Langevin equation with two fractional orders, *Physics Letters A*, 372(42): 2008, 6309–6320.
8. S. C. Lim, M. Li, and L. P. Teo, Locally self-similar fractional oscillator processes, *Fluctuation and Noise Letters*, 7(2): 2007, L169-L179.
9. M. Li, S. C. Lim, and S. Y. Chen, Exact solution of impulse response to a class of fractional oscillators and its stability, *Mathematical Problems in Engineering*, 2011: 2011, 657839, 9.
10. Y. Shao, The fractional Ornstein-Uhlenbeck process as a representation of homogeneous Eulerian velocity turbulence, *Physica D*, 83(4): 1995, 461–477.
11. P. Cheridito, H. Kawaguchi, and M. Maejima, Fractional Ornstein-Uhlenbeck processes, *Electronic Journal of Probability*, 8: 2003, 1–14.
12. M. Magdziarz, Fractional Ornstein–Uhlenbeck processes, Joseph effect in models with infinite variance, *Physica A*, 387(1): 2008, 123–133.
13. J. Gehringer and X.-M. Li, Functional limit theorems for the fractional Ornstein–Uhlenbeck process, *Journal of Theoretical Probability*, 35(1): 2022, 426–456.
14. H. G. Patel and S. N. Sharma, Some evolution equations for an Ornstein–Uhlenbeck process-driven dynamical system, *Fluctuation and Noise Letters*, 11(4): 2012, 1250020.
15. S. C. Lim and S. V. Muniandy, Generalized Ornstein-Uhlenbeck processes and associated self-similar processes, *Journal of Physics A: Mathematical and General*, 36(14): 2003, 3961–3982.
16. M. Li, *Fractional Vibrations with Applications to Euler-Bernoulli Beams*, CRC Press, Boca Raton, 2023.
17. M. Li, Analytic theory of seven classes of fractional vibrations based on elementary functions: A tutorial review, *Symmetry*, 16(9): 2024, 1202.
18. M. Li, PSD and cross PSD of responses of seven classes of fractional vibrations driven by fGn, fBm, fractional OU process, and von Kármán process, *Symmetry*, 16(5): 2024, 635.

CHAPTER 6

# Responses of Fractional Vibrations Driven by von Kármán Spectrum

THIS CHAPTER CONTRIBUTES THE analytic expressions of power spectrum density (PSD) responses and cross-PSD ones to seven classes of fractional vibrators driven by the von Kármán spectrum. It shows that there are considerable effects of orders of fractional vibrators on responses driven by the von Kármán spectrum. The responses of seven classes of fractional vibrators are of short-range dependence.

## 6.1 BACKGROUND

The von Kármán spectrum is widely used in wind engineering, ship, and ocean engineering (Faltinsen [1], Holmes [2]). Recently, seven classes of fractional vibrators were introduced in Chapter 7 in Volume I and Li [3–5]. Also, Li [3] reported the point of view that the process with the von Kármán spectrum can be taken as a type of fractional Uhlenbeck and Ornstein (UO) process with a fractal dimension of 5/3. Nonetheless, how to analytically represent the power spectrum density (PSD) and cross-PSD responses to seven classes of fractional vibrators driven by the von Kármán spectrum remains a problem unsolved. This chapter gives the analytic solution to that problem.

The rest of the chapter is organized as follows. In Sections 6.2–6.8, we put forward the analytic expressions of the PSD and cross-PSD responses

DOI: 10.1201/9781003657903-6

to seven classes of fractional vibrators excited by the von Kármán spectrum. The summary is given in Section 6.9.

## 6.2 RESPONSES OF CLASS I FRACTIONAL VIBRATORS DRIVEN BY VON KÁRMÁN SPECTRUM

### 6.2.1 Computations

Consider the motion equation of a class I fractional vibrator in the form

$$m\frac{d^{\alpha}x_1(t)}{dt^{\alpha}}+k\frac{dx_1(t)}{dt}=f(t). \tag{6.1}$$

In (5.1), $1 < \alpha < 3$, $x_1(t)$ is the response, $f(t)$ is the excitation, $m$ and $k$ are the primary mass and stiffness, respectively.

Let $f(t)$ be a process with the von Kármán spectrum in what follows. Let $S_{ff}(\omega)$ be the PSD of the von Kármán type. Then (Li [3]),

$$S_{ff}(\omega)=\frac{A_{\mathrm{vk}}}{\left[(B_{\mathrm{vk}})^2+\omega^2\right]^{5/6}}. \tag{6.2}$$

In (6.2),

$$A_{\mathrm{vk}}=\frac{4\sigma_u^2}{70.8^{5/6}\left(\dfrac{L_u^x}{U}\right)^{2/3}}$$

and

$$B_{\mathrm{vk}}=\frac{U}{70.8^{1/2}L_u^x},$$

where $L_u^x$ is turbulence integral scale, $U$ is mean speed, $u_f$ is friction velocity ($\mathrm{ms}^{-1}$), $b_v$ is friction velocity coefficient such that the variance of wind speed $\sigma_u^2=b_v u_f^2$. Figure 6.1 illustrates some plots of $S_{ff}(\omega)$. In this chapter, $L_u^x = 1$ is used simply for facilitating the illustrations. That does not lose the generality to describe the issue of the von Kármán process passing through seven classes of fractional vibration systems.

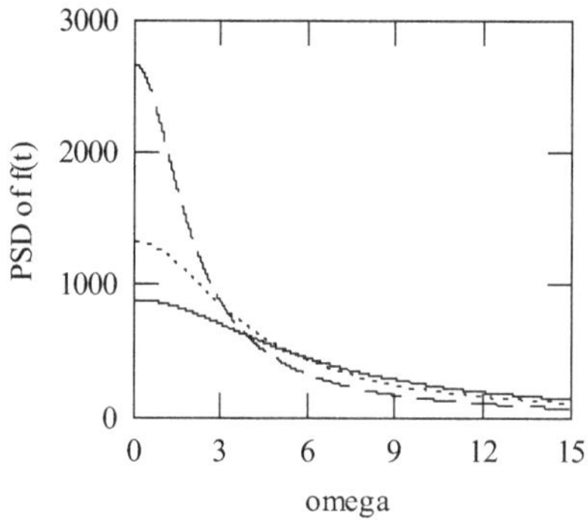


FIGURE 6.1 Plots of von Kármán spectrum with $\sigma_u = 10$ when $U = 45$ (solid), 30 (dot), 15 (dash).

**Theorem 6.1 (PSD response I)**

Denote by $S_{xx1}(\omega)$ the PSD of $x_1(t)$. Then,

$$S_{xx1}(\omega) = \frac{\dfrac{A_{\text{vk}}}{\left[(B_{\text{vk}})^2 + \omega^2\right]^{5/6}}}{k^2\left[\left(1 - \dfrac{\omega^\alpha}{\omega_n^2}\left|\cos\dfrac{\alpha\pi}{2}\right|\right)^2 + \left(\dfrac{\omega^\alpha}{\omega_n^2}\sin\dfrac{\alpha\pi}{2}\right)^2\right]}, \tag{6.3}$$

where $\omega_n^2 = \dfrac{k}{m}$.

*Proof.* Doing the operation of $S_{xx1}(\omega) = S_{ff}(\omega)|H_1(\omega)|^2$, where $H_1(\omega)$ is expressed by (Chapter 7 in Volume I or Li [3–6])

$$H_1(\omega) = \frac{1}{k\left(1 - \dfrac{\omega^\alpha}{\omega_n^2}\left|\cos\dfrac{\alpha\pi}{2}\right| + i\dfrac{\omega^\alpha}{\omega_n^2}\sin\dfrac{\alpha\pi}{2}\right)}, \tag{6.4}$$

results in (6.3). The proof is finished.

**Theorem 6.2 (cross-PSD response I)**

Let $S_{fx1}(\omega)$ be the cross-PSD between $f(t)$ and $x_1(t)$. Then,

$$S_{fx1}(\omega) = \frac{\dfrac{A_{\text{vk}}}{\left[(B_{\text{vk}})^2 + \omega^2\right]^{5/6}}}{k\left(1 - \dfrac{\omega^\alpha}{\omega_n^2}\left|\cos\dfrac{\alpha\pi}{2}\right| + i\dfrac{\omega^\alpha}{\omega_n^2}\sin\dfrac{\alpha\pi}{2}\right)}. \tag{6.5}$$

*Proof.* Using $S_{fx1}(\omega) = S_{ff}(\omega)H_1(\omega)$ with (6.4) produces (6.5). The proof ends.

Denote by $r_{xx1}(\tau)$ the autocorrelation function (ACF) of $x_1(t)$. Let $r_{ff}(\tau)$ be the ACF of $f(t)$. Let $h_1(\tau)$ be the impulse response of a class I fractional vibrator. Applying the convolution theory to $S_{xx1}(\omega) = S_{ff}(\omega)|H_1(\omega)|^2$ yields the ACF response expressed by

$$r_{xx1}(\tau) = r_{ff}(\tau) * h_1(\tau) * h_1(-\tau). \tag{6.6}$$

Similarly, applying the convolution theory to $S_{fx1}(\omega) = S_{ff}(\omega)H_1(\omega)$ produces the cross-correlation response given by

$$r_{fx1}(\tau) = r_{ff}(\tau) * h_1(\tau), \tag{6.7}$$

where (Chapter 7 in Volume I, Li [3–6])

$$h_1(t) = \frac{e^{-\frac{\omega\sin\frac{\alpha\pi}{2}}{2\left|\cos\frac{\alpha\pi}{2}\right|}t}\sin\left(\dfrac{\omega_n}{\sqrt{\omega^{\alpha-2}\left|\cos\dfrac{\alpha\pi}{2}\right|}}\sqrt{1 - \dfrac{\omega^{2\alpha}\sin^2\dfrac{\alpha\pi}{2}}{4\omega_n^2\left|\cos\dfrac{\alpha\pi}{2}\right|}}\,t\right)}{m\omega_n\sqrt{\omega^{\alpha-2}\left|\cos\dfrac{\alpha\pi}{2}\right|}\sqrt{1 - \dfrac{\omega^{2\alpha}\sin^2\dfrac{\alpha\pi}{2}}{4\omega_n^2\left|\cos\dfrac{\alpha\pi}{2}\right|}}}u(t). \tag{6.8}$$

In (6.8), $u(t)$ is the unit step function.

### 6.2.2 Effect of $\alpha$ on Responses

Figure 6.2 indicates some plots of $|H_1(\omega)|^2$. Some plots of $S_{xx1}(\omega)$ are shown in Figure 6.3. When $\alpha = 2$, a class I fractional vibrator reduces

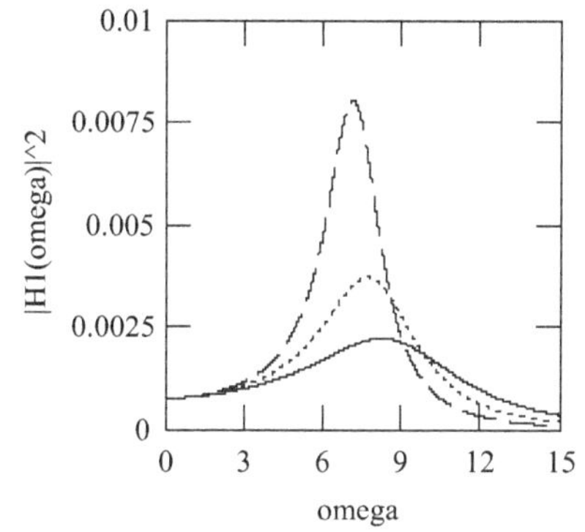


FIGURE 6.2 Plots of $|H_1(\omega)|^2$ with $\alpha = 1.6$ (solid), 1.7 (dot), 1.8 (dash) when $m = 1$ and $k = 36$.

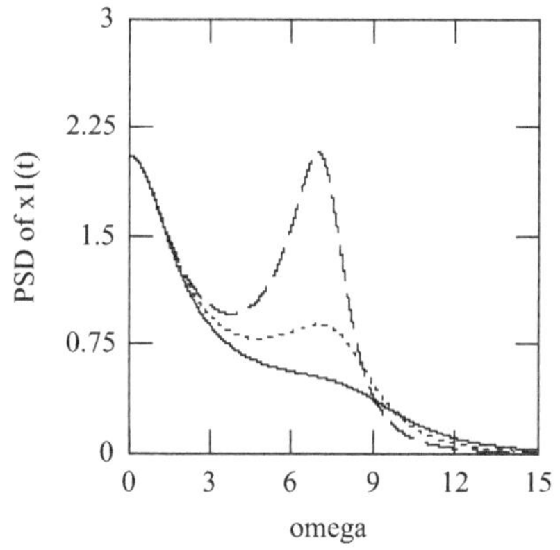


FIGURE 6.3 Plots of $S_{xx1}(\omega)$ for $m = 1$, $k = 36$, and $U = 15$ when $\alpha = 1.6$ (solid), 1.7 (dot), 1.8 (dash).

to be a conventional damping-free vibrator. In general, there are three parameters affecting response $S_{xx1}(\omega)$. From a view of system, the parameter is the fractional order $\alpha$. From the view of engineering, the parameter $U$ is crucial to $S_{xx1}(\omega)$. Figure 6.4 shows the effect of $U$ on $S_{xx1}(\omega)$. In Figure 6.5, we illustrate the plots of $|S_{fx1}(\omega)|$. The plots of the driven signal with the von Kármán spectrum and response ones are indicated in Figure 6.6. Figures 6.4–6.6 show that there is noticeable effect of $\alpha$ on the responses to class I fractional vibration systems driven by the von Kármán spectrum.

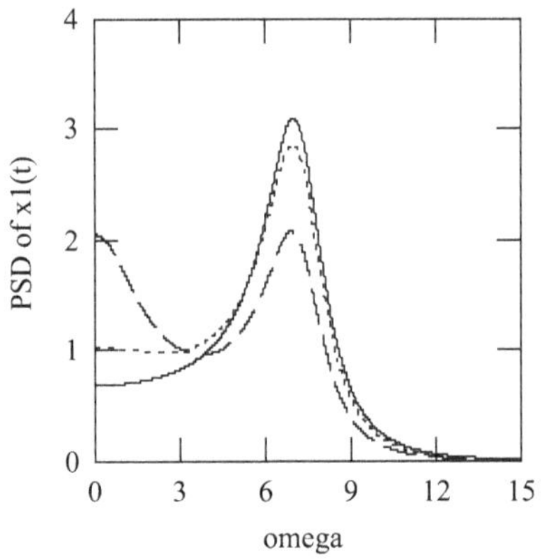


FIGURE 6.4 Observing $U$ effect on $S_{xx1}(\omega)$ when $\alpha = 1.8$, $m = 1$, $k = 36$, for $U = 40$ (solid), 30 (dot), 15 (dash).

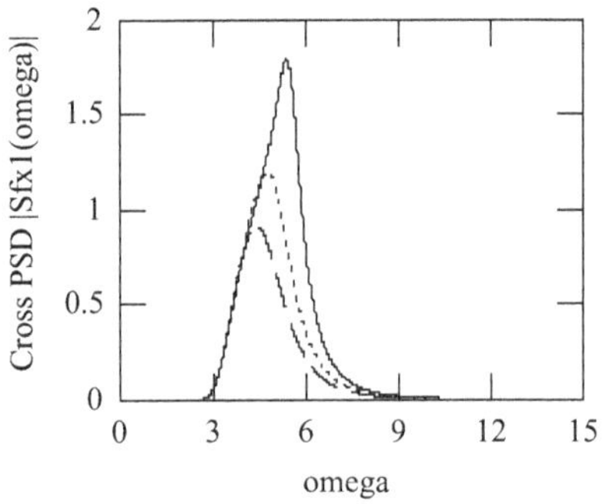


FIGURE 6.5 Illustrations of $|S_{fx1}(\omega)|$ for $m = 1$, $k = 36$, and $U = 20$ when $\alpha = 2.1$ (solid), 2.2 (dot), 2.3 (dash).

## 6.3 RESPONSES OF CLASS II FRACTIONAL VIBRATION SYSTEMS DRIVEN BY VON KÁRMÁN SPECTRUM

### 6.3.1 Computation Methods

For a class II fractional vibrator, its motion equation is given by

$$m\frac{d^2 x_2(t)}{dt^2} + c\frac{d^\beta x_2(t)}{dt^\beta} + k\frac{dx_2(t)}{dt} = f(t). \tag{6.9}$$

In (5.9), $0 < \beta < 2$, $x_2(t)$ is the response and $c$ is the primary damping.

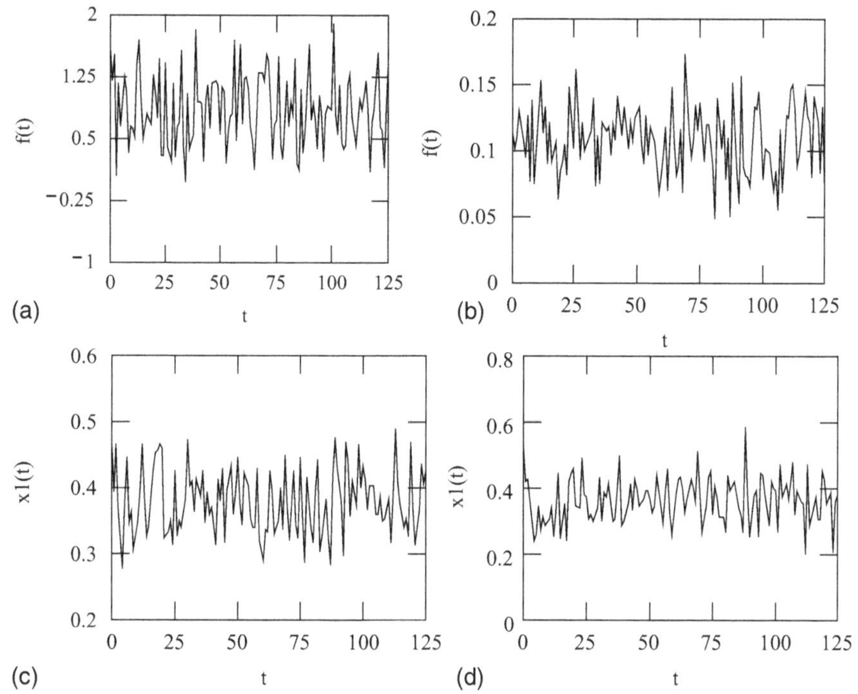


FIGURE 6.6 Plots of driven signal with the von Kármán spectrum and response series when $m = 1$, $k = 36$ and $\alpha = 1.8$. (a). Driven signal for $U = 45$. (b). Driven signal for $U = 5$. (c). Response $x_1(t)$ for $(\alpha, U) = (1.2, 20)$. (d). Response $x_1(t)$ for $(\alpha, U) = (1.8, 20)$.

## Theorem 6.3 (PSD response II)

Let $S_{xx2}(\omega)$ be the PSD of the response $x_2(t)$. Then,

$$S_{xx2}(\omega) = \frac{\dfrac{A_{\text{vk}}}{\left[(B_{\text{vk}})^2 + \omega^2\right]^{5/6}}}{k^2 \left\{ \left[ 1 - \gamma^2 \left( 1 - \dfrac{c}{m} \omega^{\beta - 2} \cos \dfrac{\beta \pi}{2} \right) \right]^2 + \left( \dfrac{2 \varsigma \omega^{\beta}}{\omega_n} \sin \dfrac{\beta \pi}{2} \right)^2 \right\}}, \tag{6.10}$$

where $\gamma = \dfrac{\omega}{\omega_n}$.

*Proof.* Using $S_{xx2}(\omega) = S_{ff}(\omega)|H_2(\omega)|^2$, where $H_2(\omega)$ is (Chapter 7 in Volume I or Li [3–6])

$$H_2(\omega)=\frac{1/k}{1-\gamma^2\left(1-\frac{c}{m}\omega^{\beta-2}\cos\frac{\beta\pi}{2}\right)+i\frac{2\varsigma\omega^{\beta}\sin\frac{\beta\pi}{2}}{\omega_n}},\tag{6.11}$$

we have (6.10). This finishes the proof.

**Theorem 6.4 (cross-PSD response II)**

Denote by $S_{fx2}(\omega)$ the cross-PSD between $f(t)$ and $x_2(t)$. Then,

$$S_{fx2}(\omega)=\frac{\frac{A_{\mathrm{vk}}}{\left[(B_{\mathrm{vk}})^2+\omega^2\right]^{5/6}}}{k\left[1-\gamma^2\left(1-2\varsigma\omega_n\omega^{\beta-2}\cos\frac{\beta\pi}{2}\right)+i\frac{2\varsigma\omega^{\beta}}{\omega_n}\sin\frac{\beta\pi}{2}\right]}.\tag{6.12}$$

*Proof.* Using $S_{fx2}(\omega)=S_{ff}(\omega)H_2(\omega)$ with (6.11) produces (6.12). The proof ends.

Denote by $h_2(t)$ the impulse response of a class II fractional vibrator. Let $r_{xx2}(\tau)$ be the ACF of $x_2(t)$. Denote by $r_{fx2}(\tau)$ the cross-correlation between $f(t)$ and $x_2(t)$. Applying the Wiener-Khinchin relation to $S_{xx2}(\omega)=S_{ff}(\omega)|H_2(\omega)|^2$ results in the ACF response $r_{xx2}(\tau)$ in the form

$$r_{xx2}(\tau)=r_{ff}(\tau)*h_2(\tau)*h_2(-\tau).\tag{6.13}$$

Similarly, applying the Wiener-Lee relation to $S_{fx2}(\omega)=S_{ff}(\omega)H_2(\omega)$ yields the cross-correlation response $r_{fx2}(\tau)$ in the form

$$r_{fx2}(\tau)=r_{ff}(\tau)*h_2(\tau).\tag{6.14}$$

In (6.13) and (6.14), according to Chapter 7 in Volume I or Li [3–6], $h_2(t)$ is expressed by

$$h_2(t)=\frac{e^{-\frac{\varsigma\omega_n\omega^{\beta-1}\sin\frac{\beta\pi}{2}}{1-\frac{c}{m}\omega^{\beta-2}\cos\frac{\beta\pi}{2}}t}\sin\frac{\omega_n\sqrt{1-\frac{\varsigma^2\omega^{2(\beta-1)}\sin^2\frac{\beta\pi}{2}}{1-\frac{c}{m}\omega^{\beta-2}\cos\frac{\beta\pi}{2}}}}{\sqrt{1-\frac{c}{m}\omega^{\beta-2}\cos\frac{\beta\pi}{2}}}t}{\omega_n m\sqrt{1-\frac{c}{m}\omega^{\beta-2}\cos\frac{\beta\pi}{2}}\sqrt{1-\frac{\varsigma^2\omega^{2(\beta-1)}\sin^2\frac{\beta\pi}{2}}{1-\frac{c}{m}\omega^{\beta-2}\cos\frac{\beta\pi}{2}}}}u(t).\tag{6.15}$$

### 6.3.2 Effect of $\beta$ on Responses

Some plots of $|H_2(\omega)|^2$ are indicated in Figure 6.7. Figure 6.8 shows some plots of $S_{xx2}(\omega)$. Figure 6.9 shows the $U$ effect on $S_{xx2}(\omega)$. Figure 6.10 illustrates some plots of $|S_{fx2}(\omega)|$. Figures 6.9 and 6.10 exhibit that there is effect

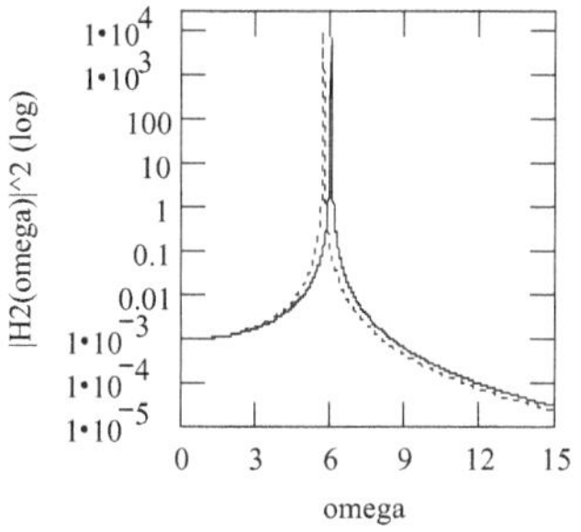


FIGURE 6.7 Plots of $|H_2(\omega)|^2$ (log) with $\beta = 0.001$ (solid), 1.999 (dot), when $m = 1$, $c = 0.1$ and $k = 36$.

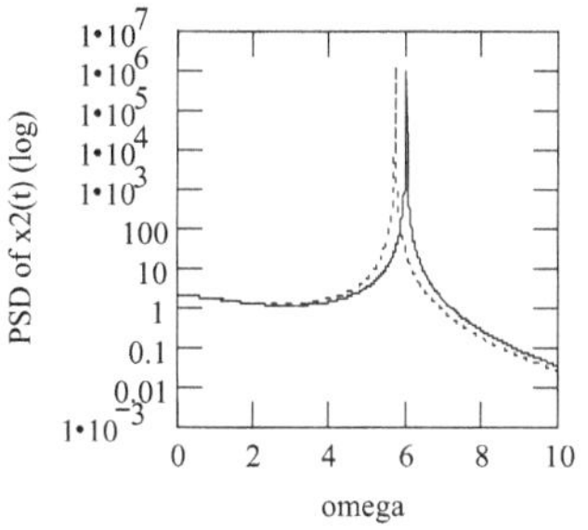


FIGURE 6.8 Plots of $S_{xx2}(\omega)$ (log) with $\beta = 0.001$ (solid), 1.999 (dot) when $m = 1$, $c = 0.1$, $k = 36$ and $U = 15$.

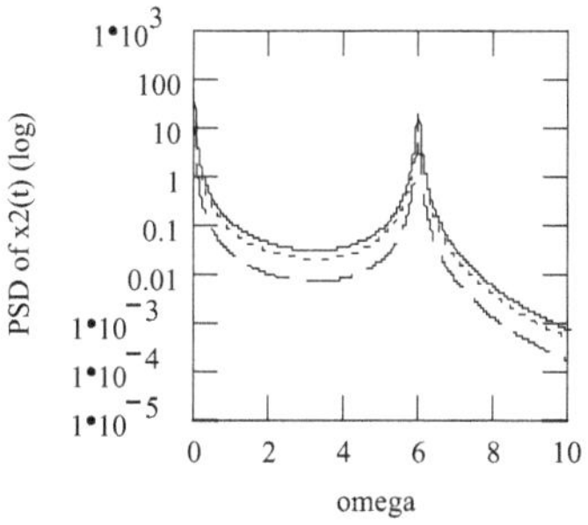


FIGURE 6.9 Observing of $U$ effect on $S_{xx2}(\omega)$ with $\beta = 1$, $m = 1$, $c = 0.1$, $k = 36$ for $U = 45$ (solid), 25 (dot), 5 (dash).

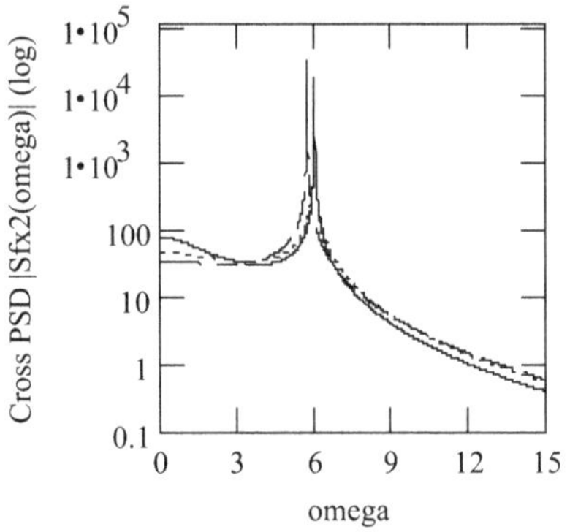


FIGURE 6.10 Plots of $|S_{fx2}(\omega)|$ (log) when $m = 1$, $c = 0.1$, and $k = 36$ with $(\beta, U) = (0.001, 15)$ (solid), $(\beta, U) = (1, 25)$ (dot), $(\beta, U) = (1.999, 35)$ (dash).

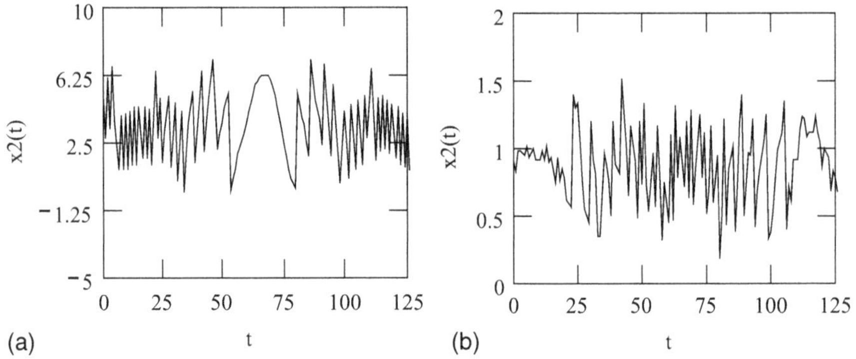


FIGURE 6.11 Response $x_2(t)$ when $m = 1$, $c = 0.1$, $k = 36$, and $U = 35$. (a). $x_2(t)$ for $\beta = 0.4$. (b). $x_2(t)$ for $\beta = 1.4$.

of $\beta$ on the responses to class II fractional vibration systems driven by the von Kármán spectrum. We also use Figure 6.11 to exhibit the effect of $\beta$ on the fluctuation range of $x_2(t)$.

## 6.4 RESPONSES OF CLASS III FRACTIONAL VIBRATORS DRIVEN BY VON KÁRMÁN SPECTRUM

### 6.4.1 Computations

The motion equation of a class III fractional vibrator is given by

$$m\frac{d^{\alpha}x_3(t)}{dt^{\alpha}} + c\frac{d^{\beta}x_3(t)}{dt^{\beta}} + kx_3(t) = f(t). \tag{6.16}$$

In (6.16), $x_3(t)$ is the response of a class III fractional vibrator.

**Theorem 6.5 (PSD response III)**

Let $S_{xx3}(\omega)$ be the PSD of $x_3(t)$. Then,

$$S_{xx3}(\omega)=\frac{\dfrac{A_{\mathrm{vk}}}{\left[(B_{\mathrm{vk}})^2+\omega^2\right]^{5/6}}}{k^2\left\{\left[1-\gamma^2\left(\omega^{\alpha-2}\left|\cos\dfrac{\alpha\pi}{2}\right|-2\varsigma\omega_n\omega^{\beta-2}\cos\dfrac{\beta\pi}{2}\right)\right]^2+\left[\dfrac{\gamma\left(\omega^{\alpha-1}\sin\dfrac{\alpha\pi}{2}+2\varsigma\omega_n\omega^{\beta-1}\sin\dfrac{\beta\pi}{2}\right)}{\omega_n\left(\omega^{\alpha-2}\left|\cos\dfrac{\alpha\pi}{2}\right|-2\varsigma\omega_n\omega^{\beta-2}\cos\dfrac{\beta\pi}{2}\right)}\right]^2\right\}}. \tag{6.17}$$

*Proof.* Considering $S_{xx3}(\omega) = S_{ff}(\omega)|H_3(\omega)|^2$, where $H_3(\omega)$ is given by (Chapter 7 in Volume I or Li [3–6])

$$H_3(\omega)=\frac{1/k}{1-\gamma^2\left(\omega^{\alpha-2}\left|\cos\dfrac{\alpha\pi}{2}\right|-2\varsigma\omega_n\omega^{\beta-2}\cos\dfrac{\beta\pi}{2}\right)+i\dfrac{\gamma\left(\omega^{\alpha-1}\sin\dfrac{\alpha\pi}{2}+2\varsigma\omega_n\omega^{\beta-1}\sin\dfrac{\beta\pi}{2}\right)}{\omega_n\left(\omega^{\alpha-2}\left|\cos\dfrac{\alpha\pi}{2}\right|-2\varsigma\omega_n\omega^{\beta-2}\cos\dfrac{\beta\pi}{2}\right)}}, \tag{6.18}$$

we have (6.17). The proof ends.

**Theorem 6.6 (cross-PSD response III)**

Let $S_{fx3}(\omega)$ be the cross-PSD between $f(t)$ and $x_3(t)$. Then,

$$S_{fx3}(\omega)=\frac{\dfrac{A_{\mathrm{vk}}}{\left[(B_{\mathrm{vk}})^2+\omega^2\right]^{5/6}}}{1-\gamma^2\left(\omega^{\alpha-2}\left|\cos\dfrac{\alpha\pi}{2}\right|-2\varsigma\omega_n\omega^{\beta-2}\cos\dfrac{\beta\pi}{2}\right)+i\dfrac{\gamma\left(\omega^{\alpha-1}\sin\dfrac{\alpha\pi}{2}+2\varsigma\omega_n\omega^{\beta-1}\sin\dfrac{\beta\pi}{2}\right)}{\omega_n\left(\omega^{\alpha-2}\left|\cos\dfrac{\alpha\pi}{2}\right|-2\varsigma\omega_n\omega^{\beta-2}\cos\dfrac{\beta\pi}{2}\right)}}. \tag{6.19}$$

*Proof.* Doing the operation of $S_{fx3}(\omega) = S_{ff}(\omega)H_3(\omega)$ with (6.18) results in (6.19). The proof ends.

Denote by $h_3(t)$ the impulse response of a class III fractional vibrator. Let $r_{xx3}(\tau)$ be the ACF of $x_3(t)$. Denote by $r_{fx3}(\tau)$ the cross-correlation between $f(t)$ and $x_3(t)$. According to the Wiener-Khinchin relation and the Wiener-Lee relation, we have the ACF response $r_{xx3}(\tau)$ in the form

$$r_{xx3}(\tau) = r_{ff}(\tau) * h_3(\tau) * h_3(-\tau), \tag{6.20}$$

and the cross-correlation response $r_{fx3}(\tau)$ in the form

$$r_{fx3}(\tau) = r_{ff}(\tau) * h_3(\tau). \tag{6.21}$$

In (6.20) and (6.21), $h_3(\tau)$ is the impulse response function of a class III fractional vibrator. According to Chapter 7 in Volume I or Li [3–6],

$$h_3(t) = \frac{e^{-\frac{m\omega^{\alpha-1}\sin\frac{\alpha\pi}{2} + c\omega^{\beta-1}\sin\frac{\beta\pi}{2}}{2\sqrt{-\left(m\omega^{\alpha-2}\cos\frac{\alpha\pi}{2} + c\omega^{\beta-2}\cos\frac{\beta\pi}{2}\right)k}}\omega_{\text{eqn}3}t} \sin\omega_{\text{eqd}3}t}{-\left(m\omega^{\alpha-2}\cos\frac{\alpha\pi}{2} + c\omega^{\beta-2}\cos\frac{\beta\pi}{2}\right)\omega_{\text{eqd}3}} u(t). \tag{6.22}$$

Refer to Chapter 7 in Volume I or Li [3–6] for the expressions of $\omega_{\text{eqn}3}$ and $\omega_{\text{eqd}3}$ in (6.22).

### 6.4.2 Effect of $(\alpha, \beta)$ on Responses

Some plots of $|H_3(\omega)|^2$ are shown in Figure 6.12. Figure 6.13 indicates some plots of $S_{xx3}(\omega)$. Some plots of $|S_{fx3}(\omega)|$ are indicated in Figure 6.14. Figures 6.13 and 6.14 exhibit that the effect of $(\alpha, \beta)$ on the responses of class III fractional vibrators driven by the von Kármán spectrum is significant.

## 6.5 RESPONSES OF CLASS IV FRACTIONAL VIBRATION SYSTEMS DRIVEN BY VON KÁRMÁN SPECTRUM

### 6.5.1 Computations

The motion equation of a class IV fractional vibrator is given by

$$m\frac{d^\alpha x_4(t)}{dt^\alpha} + k\frac{d^\lambda x_4(t)}{dt^\lambda} = f(t). \tag{6.23}$$

In (4.23), $0 \le \lambda < 1$ and $x_4(t)$ is the response of a class IV fractional vibrator.

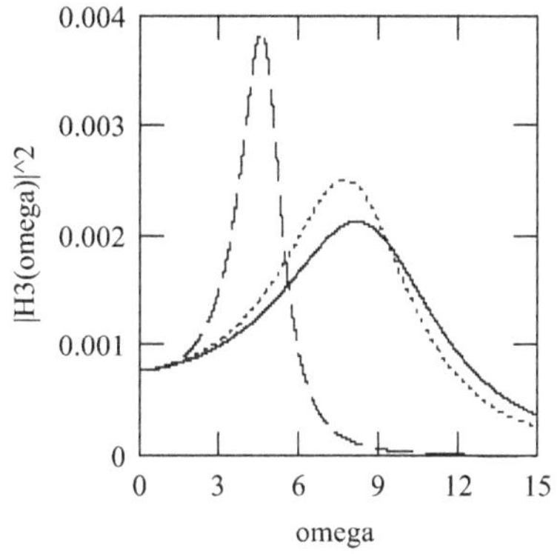


FIGURE 6.12 Plots of $|H_3(\omega)|^2$ when $m = 1, c = 0.1$, and $k = 36$, for $(\alpha, \beta) = (1.6, 0.8)$ (solid), (1.6, 1.8) (dot), (2.3, 0.8) (dash).

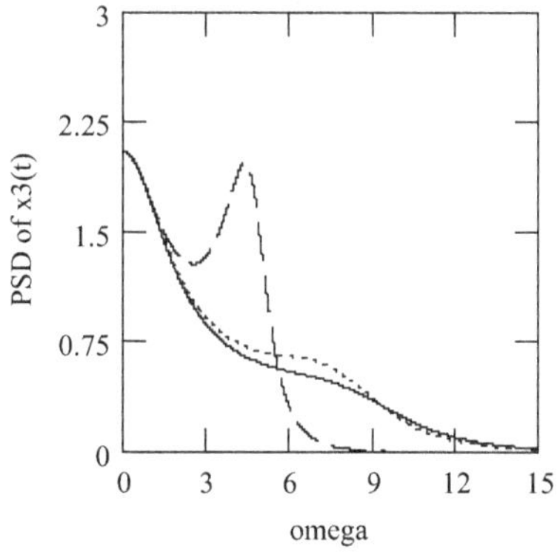


FIGURE 6.13 Plots of $S_{xx3}(\omega)$ when $m = 1, c = 0.1, k = 36$, and $U = 15$ for $(\alpha, \beta) =$ (1.6, 0.8) (solid), (1.6, 1.8) (dot), (2.3, 0.8) (dash).

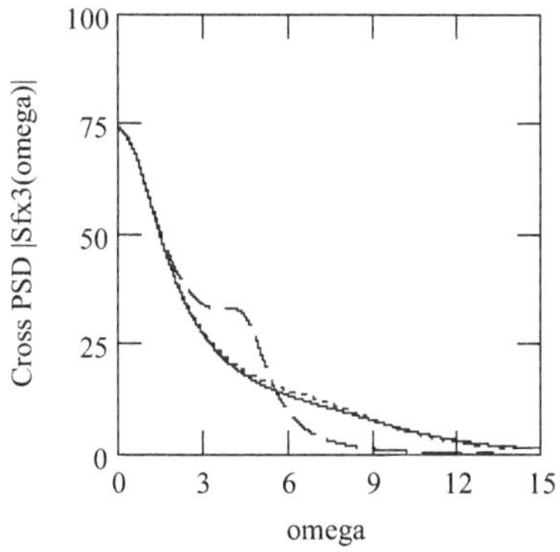


FIGURE 6.14 Plots of $|S_{fx3}(\omega)|$ when $m = 1$, $c = 0.1$, $k = 36$, and $U = 15$ for $(\alpha, \beta) = (1.6, 0.8)$ (solid), (1.6, 1.8) (dot), (2.3, 0.8) (dash).

**Theorem 6.7 (PSD response IV)**

Let $S_{xx4}(\omega)$ be the PSD of the response $x_4(t)$. Then,

$$S_{xx4}(\omega)=\frac{\dfrac{A_{\text{vk}}}{\left[(B_{\text{vk}})^2+\omega^2\right]^{5/6}}}{k^2\omega^{2\lambda}\cos^2\dfrac{\lambda\pi}{2}\left[\left(1-\gamma^2\dfrac{-\omega^{\alpha-2}\cos\dfrac{\alpha\pi}{2}}{\omega^{\lambda}\cos\dfrac{\lambda\pi}{2}}\right)^2+4\left(\gamma\dfrac{m\omega^{\alpha-1}\sin\dfrac{\alpha\pi}{2}+k\omega^{\lambda-1}\sin\dfrac{\lambda\pi}{2}}{2\sqrt{mk\omega^{\alpha+\lambda-2}\left|\cos\dfrac{\alpha\pi}{2}\right|\cos\dfrac{\lambda\pi}{2}}\sqrt{\dfrac{-\omega^{\alpha-2}\cos\dfrac{\alpha\pi}{2}}{\omega^{\lambda}\cos\dfrac{\lambda\pi}{2}}}}\right)^2\right]}. \tag{6.24}$$

*Proof.* Doing the operation of $S_{xx4}(\omega)=S_{ff}(\omega)|H_4(\omega)|^2$, where $H_4(\omega)$ is given by (Chapter 7 in Volume I or Li [3, 4] or Li [6])

$$H_4(\omega)=\frac{1}{k\omega^{\lambda}\cos\dfrac{\lambda\pi}{2}\left(1-\gamma^2\dfrac{-\omega^{\alpha-2}\cos\dfrac{\alpha\pi}{2}}{\omega^{\lambda}\cos\dfrac{\lambda\pi}{2}}+i2\gamma\dfrac{m\omega^{\alpha-1}\sin\dfrac{\alpha\pi}{2}+k\omega^{\lambda-1}\sin\dfrac{\lambda\pi}{2}}{2\sqrt{mk\omega^{\alpha+\lambda-2}\left|\cos\dfrac{\alpha\pi}{2}\right|\cos\dfrac{\lambda\pi}{2}}\sqrt{\dfrac{-\omega^{\alpha-2}\cos\dfrac{\alpha\pi}{2}}{\omega^{\lambda}\cos\dfrac{\lambda\pi}{2}}}}\right)}, \tag{6.25}$$

produces (6.24). This finishes the proof.

**Theorem 6.8 (cross-PSD response IV)**

Denote by $S_{fx4}(\omega)$ the cross PSD between $f(t)$ and $x_4(t)$. Then,

$$S_{fx4}(\omega)=\frac{\dfrac{A_{\text{vk}}}{\left[(B_{\text{vk}})^2+\omega^2\right]^{5/6}}}{k\omega^{\lambda}\cos\dfrac{\lambda\pi}{2}\left(1-\gamma^2\dfrac{-\omega^{\alpha-2}\cos\dfrac{\alpha\pi}{2}}{\omega^{\lambda}\cos\dfrac{\lambda\pi}{2}}+i2\gamma\dfrac{m\omega^{\alpha-1}\sin\dfrac{\alpha\pi}{2}+k\omega^{\lambda-1}\sin\dfrac{\lambda\pi}{2}}{2\sqrt{mk\omega^{\alpha+\lambda-2}\left|\cos\dfrac{\alpha\pi}{2}\right|\cos\dfrac{\lambda\pi}{2}}}\sqrt{\dfrac{-\omega^{\alpha-2}\cos\dfrac{\alpha\pi}{2}}{\omega^{\lambda}\cos\dfrac{\lambda\pi}{2}}}\right)}. \tag{6.26}$$

*Proof.* Doing the operation of $S_{fx4}(\omega)=S_{ff}(\omega)H_4(\omega)$ and considering (6.25) produces (6.26). The proof is finished.

In the time domain, we have the ACF response

$$r_{xx4}(\tau)=r_{ff}(\tau)*h_4(\tau)*h_4(-\tau), \tag{6.27}$$

where $r_{xx4}(\tau)$ is the ACF of $x_4(t)$ and $h_4(\tau)$ is the impulse response of a class IV fractional vibrator. In addition, the cross-correlation response is

$$r_{fx4}(\tau)=r_{ff}(\tau)*h_4(\tau), \tag{6.28}$$

where $r_{fx4}(\tau)$ is the cross-correlation between the excitation $f(t)$ and $x_4(t)$. In (6.27) and (6.28) (Chapter 7 in Volume I or Li [3, 4] or Li [6]),

$$h_4(t)=e^{-\frac{m\omega^{\alpha-1}\sin\frac{\alpha\pi}{2}+k\omega^{\lambda-1}\sin\frac{\lambda\pi}{2}}{2\sqrt{mk\omega^{\alpha+\lambda-2}\left|\cos\frac{\alpha\pi}{2}\right|\cos\frac{\lambda\pi}{2}}}\sqrt{\frac{\omega^{\lambda}\cos\frac{\lambda\pi}{2}}{-\omega^{\alpha-2}\cos\frac{\alpha\pi}{2}}}\omega_n t}\frac{1}{m_{\text{eq4}}\omega_{\text{eqd4}}}\sin\omega_{\text{eqd4}}tu(t). \tag{6.29}$$

Refer Chapter 7 in Volume I or Li [3, 4] or Li [6] for the expressions of $m_{eq4}$, $\omega_{eqn4}$, and $\omega_{eqd4}$ in (6.29).

### 6.5.2 Effect of $(\alpha, \lambda)$ on Responses

We illustrate some plots of $|H_4(\omega)|^2$ in Figure 6.15. Figure 6.16 indicates some plots of $S_{xx4}(\omega)$. Some plots of $|S_{fx4}(\omega)|$ are indicated in Figure 6.17. Figures 6.16 and 6.17 demonstrate that the effect of $(\alpha, \lambda)$ on the responses of class IV fractional vibrators driven by the von Kármán spectrum is noticeable.

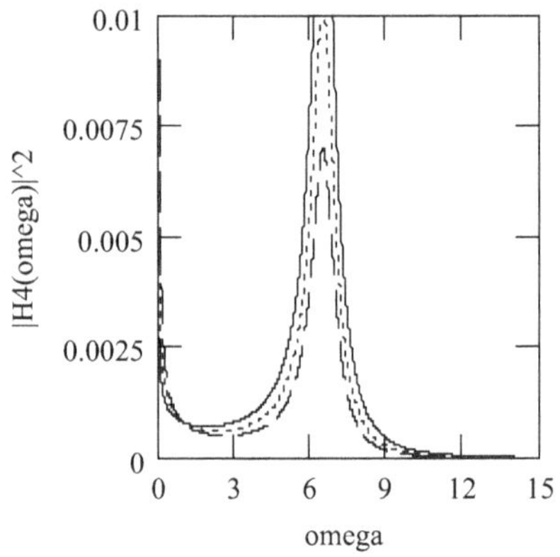


FIGURE 6.15 Plots of $|H_4(\omega)|^2$ when $m = 1$, $c = 0$, and $k = 36$, for $(\alpha, \lambda) = (2.1, 0.2)$ (solid), (2.2, 0.3) (dot), (2.3, 0.4) (dash).

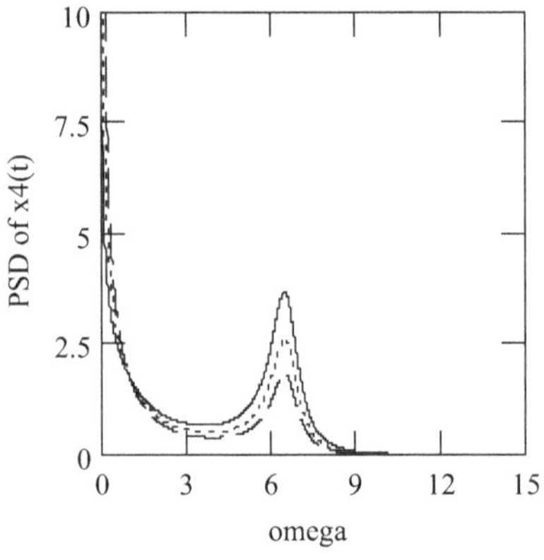


FIGURE 6.16 Plots of $S_{xx4}(\omega)$ when $m = 1$, $c = 0$, $k = 36$, and $U = 15$ for $(\alpha, \lambda) =$ (2.1, 0.2) (solid), (2.2, 0.3) (dot), (2.3, 0.4) (dash).

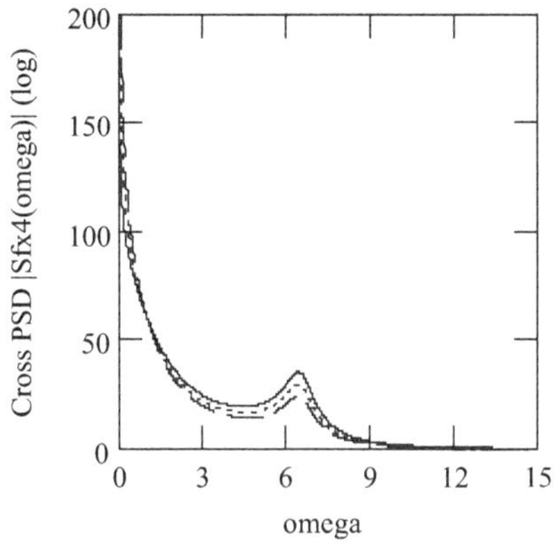


FIGURE 6.17 Plots of $|S_{fx4}(\omega)|$ when $m = 1$, $c = 0$, $k = 36$, and $U = 15$ for $(\alpha, \lambda) =$ (2.1, 0.2) (solid), (2.2, 0.3) (dot), (2.3, 0.4) (dash).

## 6.6 RESPONSES OF CLASS V FRACTIONAL VIBRATORS DRIVEN BY VON KÁRMÁN SPECTRUM

### 6.6.1 Computation Methods

Consider the motion equation of a class V fractional vibrator

$$m\frac{d^2x_5(t)}{dt^2} + k\frac{d^\lambda x_5(t)}{dt^\lambda} = f(t). \tag{6.30}$$

In (6.30), $x_5(t)$ is the response of a class V fractional vibrator.

**Theorem 6.9 (PSD response V)**

Denote by $S_{xx5}(\omega)$ the PSD of the response $x_5(t)$. Then,

$$S_{xx5}(\omega) = \frac{\dfrac{A_{\text{vk}}}{\left[(B_{\text{vk}})^2 + \omega^2\right]^{5/6}}}{k^2\omega^{2\lambda}\cos^2\dfrac{\lambda\pi}{2}\left[\left(1 - \dfrac{\gamma^2}{\omega^\lambda \cos\dfrac{\lambda\pi}{2}}\right)^2 + 4\gamma^2\left(\dfrac{k\omega^{\lambda-1}\sin\dfrac{\lambda\pi}{2}}{2\sqrt{mk\omega^\lambda\cos\dfrac{\lambda\pi}{2}}}\sqrt{\dfrac{1}{\omega^\lambda\cos\dfrac{\lambda\pi}{2}}}\right)^2\right]}. \tag{6.31}$$

*Proof.* Doing the operation of $S_{xx5}(\omega) = S_{ff}(\omega)|H_5(\omega)|^2$, where $H_5(\omega)$ is expressed by (Chapter 7 in Volume I or Li [3, 4] or Li [6])

$$H_5(\omega)=\frac{1}{k\omega^{\lambda}\cos\frac{\lambda\pi}{2}\left(1-\frac{\gamma^2}{\omega^{\lambda}\cos\frac{\lambda\pi}{2}}+i2\gamma\frac{k\omega^{\lambda-1}\sin\frac{\lambda\pi}{2}}{2\sqrt{mk\omega^{\lambda}\cos\frac{\lambda\pi}{2}}}\sqrt{\frac{1}{\omega^{\lambda}\cos\frac{\lambda\pi}{2}}}\right)}, \tag{6.32}$$

results in (6.31). The proof ends.

**Theorem 6.10 (cross-PSD response V)**

Let $S_{fx5}(\omega)$ be the cross-PSD between $f(t)$ and $x_5(t)$. Then,

$$S_{fx5}(\omega)=\frac{\frac{A_{\mathrm{vk}}}{\left[(B_{\mathrm{vk}})^2+\omega^2\right]^{5/6}}}{k\omega^{\lambda}\cos\frac{\lambda\pi}{2}\left(1-\frac{\gamma^2}{\omega^{\lambda}\cos\frac{\lambda\pi}{2}}+i2\gamma\frac{k\omega^{\lambda-1}\sin\frac{\lambda\pi}{2}}{2\sqrt{mk\omega^{\lambda}\cos\frac{\lambda\pi}{2}}}\sqrt{\frac{1}{\omega^{\lambda}\cos\frac{\lambda\pi}{2}}}\right)}. \tag{6.33}$$

*Proof.* Doing the operation of $S_{fx5}(\omega) = S_{ff}(\omega)H_5(\omega)$ and taking into account (6.32) yields (6.33). This finishes the proof.

Let $h_5(t)$ be the impulse response of a class V fractional vibrator. Then, we have

$$r_{xx5}(\tau) = r_{ff}(\tau)*h_5(\tau)*h_5(-\tau), \tag{6.34}$$

where $r_{xx5}(\tau)$ is the ACF of $x_5(t)$. In addition,

$$r_{fx5}(\tau) = r_{ff}(\tau)*h_5(\tau), \tag{6.35}$$

where $r_{fx5}(\tau)$ is the cross-correlation between the excitation $f(t)$ and the response $x_5(t)$. In (6.34) and (6.35) (Chapter 7 in Volume I or Li [3, 4] or Li [6]),

$$h_5(t) = e^{-\frac{k\omega^{\lambda-1}\sin\frac{\lambda\pi}{2}}{2\sqrt{mk\omega^{\lambda}\cos\frac{\lambda\pi}{2}}}\sqrt{\omega^{\lambda}\cos\frac{\lambda\pi}{2}}\omega_n t} \frac{1}{m\omega_{\mathrm{eqd5}}}\sin\omega_{\mathrm{eqd5}} t u(t). \tag{6.36}$$

Refer to Chapter 7 in Volume I or Li [3, 4] or Li [6] for the expression of $\omega_{\mathrm{eqd5}}$ in (6.36).

### 6.6.2 Effect of $\lambda$ on Responses

Figure 6.18 illustrates some plots of $|H_5(\omega)|^2$. Figure 6.19 is for some plots of $S_{xx5}(\omega)$. Figure 6.20 is for some plots of $|S_{fx5}(\omega)|$. Figures 6.19 and 6.20 illustrate the considerable effect of $\lambda$ on the responses of class V fractional vibrators driven by the von Kármán spectrum.

## 6.7 RESPONSES OF CLASS VI FRACTIONAL VIBRATORS DRIVEN BY VON KÁRMÁN SPECTRUM

### 6.7.1 Computations

Consider the motion equation of a class VI fractional vibrator

$$m\frac{d^{\alpha}x_6(t)}{dt^{\alpha}} + c\frac{d^{\beta}x_6(t)}{dt^{\beta}} + k\frac{d^{\lambda}x_6(t)}{dt^{\lambda}} = f(t). \tag{6.37}$$

In (6.37), $x_6(t)$ is the response of a class VI fractional vibrator.

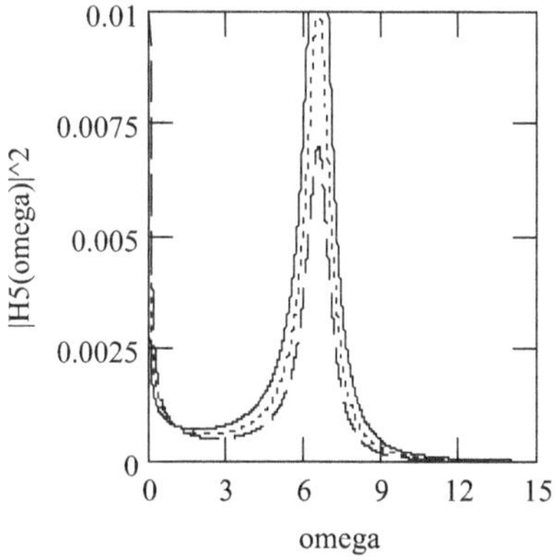


FIGURE 6.18 Plots of $|H_5(\omega)|^2$ when $m = 1$, $c = 0$, and $k = 36$, for $\lambda = 0.2$ (solid), 0.3 (dot), 0.4 (dash).

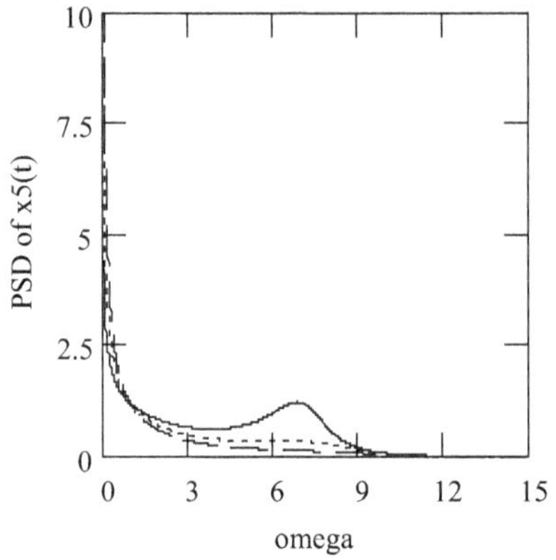


FIGURE 6.19 Plots of $S_{xx5}(\omega)$ when $m = 1$, $c = 0$, $k = 36$, and $U = 25$ for $\lambda = 0.2$ (solid), 0.3 (dot), 0.4 (dash).

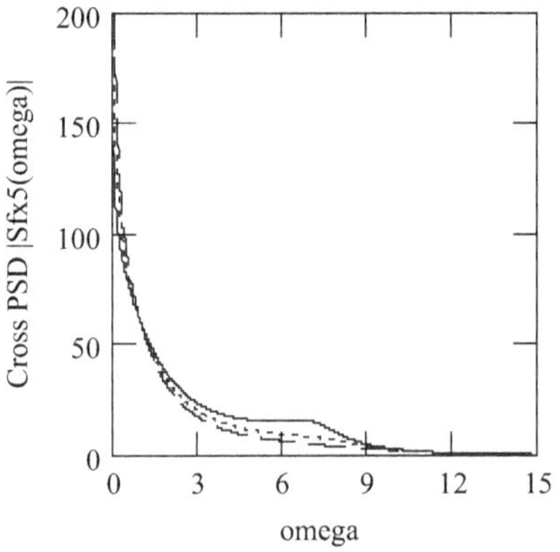


FIGURE 6.20 Plots of e $|S_{fx5}(\omega)|$ when $m = 1$, $c = 0$, $k = 36$, and $U = 25$ for $\lambda = 0.2$ (solid), 0.3 (dot), 0.4 (dash).

**Theorem 6.11 (PSD response VI)**

Let $S_{xx6}(\omega)$ be the PSD of the response $x_6(t)$. Then,

$$S_{xx6}(\omega) = \frac{1}{k^2} \frac{\dfrac{A_{\mathrm{vk}}}{\left[(B_{\mathrm{vk}})^2 + \omega^2\right]^{5/6}}}{\left[\omega^{\lambda} \cos\dfrac{\lambda\pi}{2} + \gamma^2\left(\omega^{\alpha-2}\cos\dfrac{\alpha\pi}{2} + 2\varsigma\omega_n\omega^{\beta-2}\cos\dfrac{\beta\pi}{2}\right)\right]^2 + \gamma^2\left(\omega^{\alpha-1}\sin\dfrac{\alpha\pi}{2} + 2\zeta\omega_n\omega^{\beta-1}\sin\dfrac{\beta\pi}{2} + \omega_n^2\omega^{\lambda-1}\sin\dfrac{\lambda\pi}{2}\right)^2}. \tag{6.38}$$

*Proof.* Following Chapter 7 in Volume I or Li [3, 4] or Li [6], $H_6(\omega)$ is given by

$$H_6(\omega)=\frac{1}{k\left[\begin{array}{l}\omega^{\lambda}\cos\frac{\lambda\pi}{2}+\gamma^2\left(\omega^{\alpha-2}\cos\frac{\alpha\pi}{2}+2\varsigma\omega_n\omega^{\beta-2}\cos\frac{\beta\pi}{2}\right)\\+i\gamma\left(\omega^{\alpha-1}\sin\frac{\alpha\pi}{2}+2\zeta\omega_n\omega^{\beta-1}\sin\frac{\beta\pi}{2}+\omega_n^2\omega^{\lambda-1}\sin\frac{\lambda\pi}{2}\right)\end{array}\right]}. \tag{6.39}$$

Considering $S_{xx6}(\omega) = S_{ff}(\omega)|H_6(\omega)|^2$ produces (6.38). The proof ends.

**Theorem 6.12 (cross-PSD response VI)**

Denote by $S_{fx6}(\omega)$ the cross PSD between $f(t)$ and $x_6(t)$. Then,

$$S_{fx6}(\omega)=\frac{\dfrac{A_{\mathrm{vk}}}{\left[(B_{\mathrm{vk}})^2+\omega^2\right]^{5/6}}}{k\left[\begin{array}{l}\omega^{\lambda}\cos\frac{\lambda\pi}{2}+\gamma^2\left(\omega^{\alpha-2}\cos\frac{\alpha\pi}{2}+2\varsigma\omega_n\omega^{\beta-2}\cos\frac{\beta\pi}{2}\right)\\+i\gamma\left(\omega^{\alpha-1}\sin\frac{\alpha\pi}{2}+2\zeta\omega_n\omega^{\beta-1}\sin\frac{\beta\pi}{2}+\omega_n^2\omega^{\lambda-1}\sin\frac{\lambda\pi}{2}\right)\end{array}\right]}. \tag{6.40}$$

*Proof.* Doing the operation of $S_{fx6}(\omega) = S_{ff}(\omega)H_6(\omega)$ with the consideration of (6.39) results in (6.40) (Li [7]). The proof is finished.

Let $h_6(\tau)$ be the impulse response of a class VI fractional vibrator. Applying the Wiener-Khinchin relation and the Wiener-Lee relation respectively to $S_{xx6}(\omega) = S_{ff}(\omega)|H_6(\omega)|^2$ and $S_{fx6}(\omega) = S_{ff}(\omega)H_6(\omega)$, we have

$$r_{xx6}(\tau) = r_{ff}(\tau)*h_6(\tau)*h_6(-\tau), \tag{6.41}$$

where $r_{xx6}(\tau)$ is the ACF of $x_6(t)$, and

$$r_{fx6}(\tau) = r_{ff}(\tau)*h_6(\tau), \tag{6.42}$$

where $r_{fx6}(\tau)$ is the cross-correlation between $f(t)$ and $x_6(t)$. According to Chapter 7 in Volume I or Li [3, 4] or Li [6], $h_6(t)$ is given by

$$h_6(t)=e^{-\varsigma_{\mathrm{eq6}}\omega_{\mathrm{eqn6}}t}\frac{1}{m_{\mathrm{eq6}}\omega_{\mathrm{eqd6}}}\sin\omega_{\mathrm{eqd6}}tu(t). \tag{6.43}$$

Refer to Chapter 7 in Volume I or Li [3, 4] or Li [6] for the expressions of $\zeta_{eq6}$, $m_{eq6}$, $\omega_{eqn6}$, and $\omega_{eqd6}$ in (6.43).

### 6.7.2 Effect of $(\alpha, \beta, \lambda)$ on Responses

Figure 6.21 illustrates some plots of $|H_6(\omega)|^2$. Figure 6.22 shows some plots of $S_{xx6}(\omega)$. Figure 6.23 indicates some plots of $|S_{fx6}(\omega)|$. Figures 6.22 and 6.23 show the significant effect of $(\alpha, \beta, \lambda)$ on the responses to class VI fractional vibration systems under the excitation of the von Kármán spectrum.

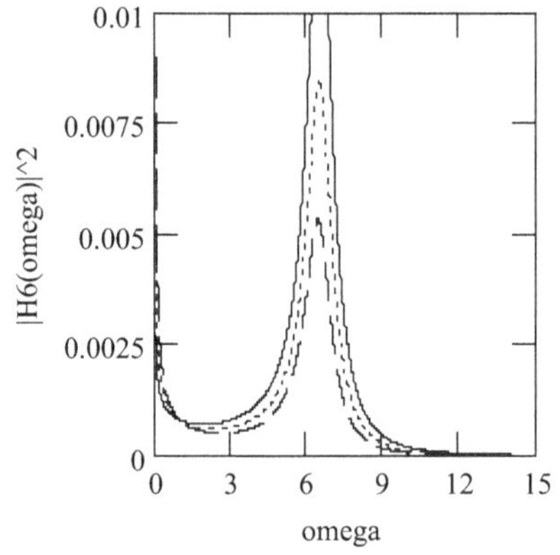


FIGURE 6.21 Plots of $|H_6(\omega)|^2$ when $m = 1$, $c = 0.1$, and $k = 36$, for $(\alpha, \beta, \lambda) = (2.1, 0.8, 0.2)$ (solid), $(2.2, 1.2, 0.4)$ (dot), $(2.3, 1.6, 0.4)$ (dash).

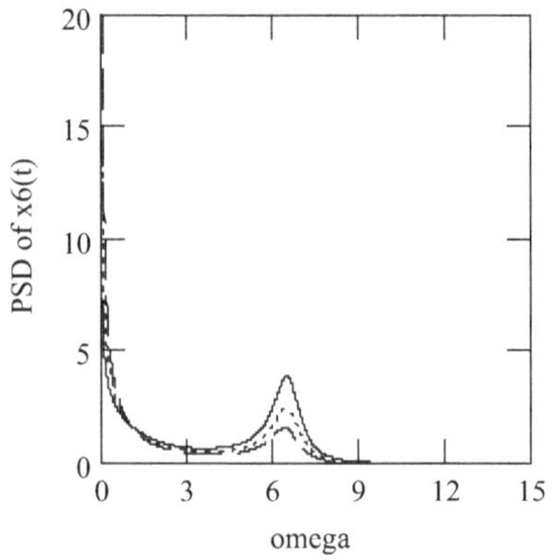


FIGURE 6.22 Plots of $S_{xx6}(\omega)$ when $m = 1$, $c = 0.1$, $k = 36$, and $U = 15$ for $(\alpha, \beta, \lambda) = (2.1, 0.8, 0.2)$ (solid), $(2.2, 1.2, 0.3)$ (dot), $(2.3, 1.6, 0.4)$ (dash).

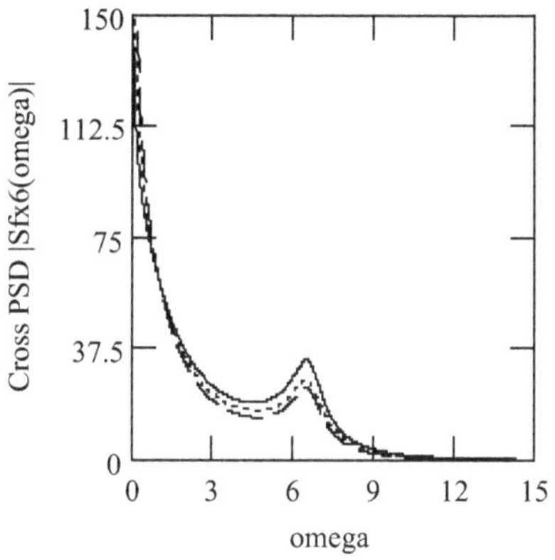


FIGURE 6.23 Plots of $|S_{fx6}(\omega)|$ when $m=1, c=0.1, k=36$, and $U=15$ for $(\alpha, \beta, \lambda)=$ (2.1, 0.8, 0.2) (solid), (2.2, 1.2, 0.3) (dot), (2.3, 1.6, 0.4) (dash).

## 6.8 RESPONSES OF CLASS VII FRACTIONAL VIBRATORS DRIVEN BY VON KÁRMÁN SPECTRUM

### 6.8.1 Computations

Consider the motion equation of a class VII fractional vibrator

$$m\frac{d^2x_7(t)}{dt^2}+c\frac{d^\beta x_7(t)}{dt^\beta}+k\frac{d^\lambda x_7(t)}{dt^\lambda}=f(t). \tag{6.44}$$

In (6.44), $x_7(t)$ is the response of a class VII fractional vibrator.

**Theorem 6.13 (PSD response VII)**

Denote by $S_{xx7}(\omega)$ the PSD of $x_7(t)$. Then,

$$S_{xx7}(\omega)=\frac{1}{k^2}\frac{\dfrac{A_{\mathrm{vk}}}{\left[(B_{\mathrm{vk}})^2+\omega^2\right]^{5/6}}}{\left[\omega^\lambda\cos\dfrac{\lambda\pi}{2}-\gamma\left(1-2\varsigma\omega_n\omega^{\beta-2}\cos\dfrac{\beta\pi}{2}\right)\right]^2+\gamma^2\left(2\varsigma\omega^{\beta-1}\sin\dfrac{\beta\pi}{2}+\omega_n\omega^{\lambda-1}\sin\dfrac{\lambda\pi}{2}\right)^2}. \tag{6.45}$$

*Proof.* Doing the operation of $S_{xx7}(\omega)=S_{ff}(\omega)|H_7(\omega)|^2$, where $H_7(\omega)$ is given by (Chapter 7 in Volume I or Li [6])

$$H_7(\omega)=\frac{1}{k\begin{bmatrix}\omega^{\lambda}\cos\frac{\lambda\pi}{2}-\gamma\left(1-2\varsigma\omega_n\omega^{\beta-2}\cos\frac{\beta\pi}{2}\right)\\+i\gamma\left(2\varsigma\omega^{\beta-1}\sin\frac{\beta\pi}{2}+\omega_n\omega^{\lambda-1}\sin\frac{\lambda\pi}{2}\right)\end{bmatrix}}, \tag{6.46}$$

yields (6.45). The proof is finished.

**Theorem 6.14 (cross-PSD response VII)**

Let $S_{fx7}(\omega)$ be the cross PSD between $f(t)$ and $x_7(t)$. Then,

$$S_{fx7}(\omega)=\frac{\dfrac{A_{\mathrm{vk}}}{\left[(B_{\mathrm{vk}})^2+\omega^2\right]^{5/6}}}{k\begin{bmatrix}\omega^{\lambda}\cos\frac{\lambda\pi}{2}-\gamma\left(1-2\varsigma\omega_n\omega^{\beta-2}\cos\frac{\beta\pi}{2}\right)\\+i\gamma\left(2\varsigma\omega^{\beta-1}\sin\frac{\beta\pi}{2}+\omega_n\omega^{\lambda-1}\sin\frac{\lambda\pi}{2}\right)\end{bmatrix}}. \tag{6.47}$$

*Proof.* Doing the operation of $S_{fx7}(\omega) = S_{ff}(\omega)H_7(\omega)$ and considering (6.46) yields (6.47). The proof ends.

Let $h_7(t)$ be the impulse response of a class VII fractional vibration system. According to the Wiener-Khinchin relation and the Wiener-Lee relation, we have

$$r_{xx7}(\tau) = r_{ff}(\tau)*h_7(\tau)*h_7(-\tau), \tag{6.48}$$

where $r_{xx7}(\tau)$ is the ACF of $x_7(t)$ and

$$r_{fx7}(\tau) = r_{ff}(\tau)*h_7(\tau), \tag{6.49}$$

where $r_{fx7}(\tau)$ is the cross-correlation between $f(t)$ and $x_7(t)$. Following Chapter 7 in Volume I or Li [6],

$$h_7(t)=e^{-\varsigma_{\mathrm{eq7}}\omega_{\mathrm{eqn7}}t}\frac{1}{m_{\mathrm{eq7}}\omega_{\mathrm{eqd7}}}\sin\omega_{\mathrm{eqd7}}t, \quad t\geq 0. \tag{6.50}$$

Refer to Chapter 7 in Volume I or Li [6] for the expressions of $\zeta_{\mathrm{eq7}}$, $m_{\mathrm{eq7}}$, $\omega_{\mathrm{eqn7}}$, and $\omega_{\mathrm{eqd7}}$ in (6.50).

### 6.8.2 Effect of ($\beta$, $\lambda$) on Responses

Figure 6.24 indicates some plots of $|H_7(\omega)|^2$. Figure 6.25 gives some plots of $S_{xx7}(\omega)$. Figure 6.26 is for some plots of |PSD $S_{fx7}(\omega)$|. Figures 6.25 and 6.26 exhibit the noticeable effect of fractional orders ($\beta$, $\lambda$) on the responses to class VII fractional vibrators under the excitation of the von Kármán spectrum.

## 6.9 SUMMARY

We have presented the closed-form analytic expressions of the PSD and cross-PSD responses to seven classes of fractional vibrators driven by the von Kármán spectrum in Theorems 6.1–6.14, respectively. We have

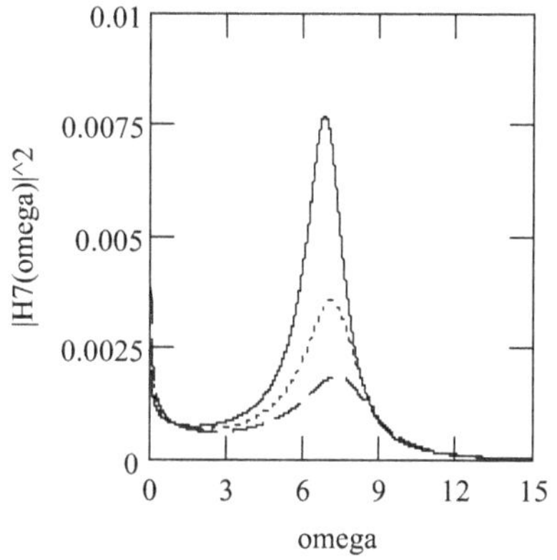


FIGURE 6.24 Plots of $|H_7(\omega)|^2$ when $m = 1, c = 0.1$, and $k = 36$, for $(\beta, \lambda) = (0.5, 0.15)$ (solid), (0.6, 0.20) (dot), (0.7, 0.25) (dash).

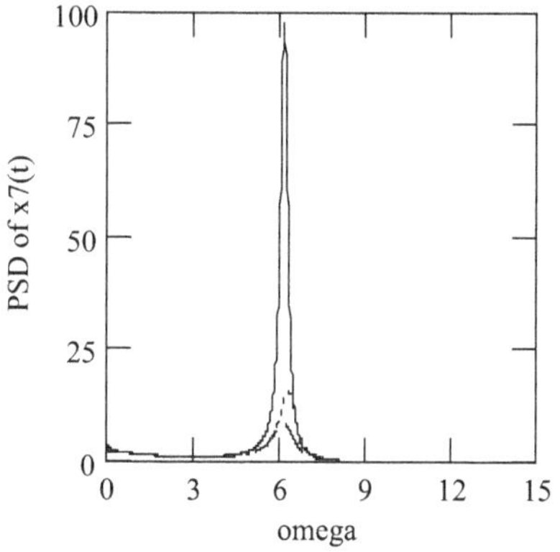


FIGURE 6.25 Plots of $S_{xx7}(\omega)$ when $m = 1, c = 0.1, k = 36$, and $U = 15$ for $(\beta, \lambda) =$ (0.01, 0.03) (solid), (1.00, 0.06) (dot), (1.99, 0.09) (dash).

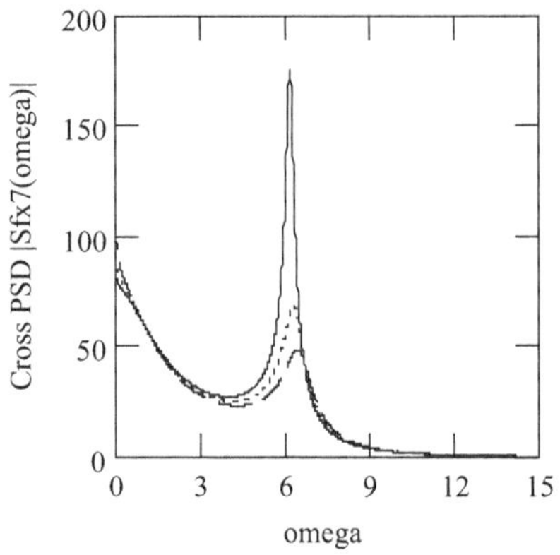


FIGURE 6.26 Plots of $|S_{fx7}(\omega)|$ when $m = 1$, $c = 0.1$, $k = 36$, and $U = 15$ for $(\beta, \lambda) = (0.01, 0.03)$ (solid), $(1.00, 0.06)$ (dot), $(1.99, 0.09)$ (dash).

demonstrated that the orders of fractional vibrators affect the responses noticeably. The responses of seven classes of fractional vibrators are of short-range dependence due to $S_{xxi}(0) < \infty$ for $i = 1, \ldots, 7$.

## 6.10 EXERCISES

6.1. Let $S_{ff}(\omega) = \dfrac{A_{vk}}{\left[(B_{vk})^2 + \omega^2\right]^{5/6}}$, where $A_{vk} = \dfrac{4\sigma_u^2}{70.8^{5/6}\left(\dfrac{L_u^x}{U}\right)^{2/3}}$,

$B_{vk} = \dfrac{U}{70.8^{1/2} L_u^x}$, $L_u^x$ is turbulence integral scale, $U$ is mean speed, $u_f$ is friction velocity (ms$^{-1}$), $b_v$ is friction velocity coefficient such that the variance of wind speed $\sigma_u^2 = b_v u_f^2$. Find $r_{ff}(\tau) = \mathrm{F}^{-1}[S_{ff}(\omega)]$.

6.2. Find the condition for $\int_{-\infty}^{\infty} r_{ff}(\tau) < \infty$, where $r_{ff}(\tau) = \mathrm{F}^{-1}[S_{ff}(\omega)]$ in Exercise 6.1.

6.3. Following Chapter 7 in Volume I, we write $h_6(t) = e^{-\varsigma_{eq6}\omega_{eqn6}t} \dfrac{1}{m_{eq6}\omega_{eqd6}}$ $\sin \omega_{eqd6} t u(t)$. Find $h_6(\tau) * h_6(-\tau)$.

6.4. Denote by $r_{ff}(\tau)$ the inverse Fourier transform of $S_{ff}(\omega)$. Find $r_{xx6}(\tau) = r_{ff}(\tau) * h_6(\tau) * h_6(-\tau)$, where $r_{ff}(\tau) = \mathrm{F}^{-1}[S_{ff}(\omega)]$ in Exercise 6.1.

6.5. Find $r_{fx6}(\tau) = r_{ff}(\tau) * h_6(\tau)$, where $r_{ff}(\tau) = \mathrm{F}^{-1}[S_{ff}(\omega)]$ in Exercise 6.1.

6.6. Find the Fourier transform of $r_{fx6}(\tau)$.

6.7. Find the condition for $\int_{-\infty}^{\infty} r_{xx6}(\tau) < \infty$.

## REFERENCES


1. O. M. Faltinsen, *Sea Loads on Ships and Offshore Structures*, 2nd ed., Cambridge University Press, Cambridge, 1990.
2. J. D. Holmes, *Wind Loading of Structure*, 2nd ed., Taylor & Francis, London and New York, 2007.
3. M. Li, *Fractional Vibrations with Applications to Euler-Bernoulli Beams*, CRC Press, Boca Raton, 2023.
4. M. Li, *Theory of Fractional Engineering Vibrations*, Walter de Gruyter, Berlin/Boston, 2021.
5. M. Li, Three classes of fractional oscillators, *Symmetry-Basel*, 10(2): 2018, 91.
6. M. Li, Analytic theory of seven classes of fractional vibrations based on elementary functions: A tutorial review, *Symmetry*, 16(9): 2024, 1202.
7. M. Li, PSD and cross PSD of responses of seven classes of fractional vibrations driven by fGn, fBm, fractional OU process, and von Kármán process, *Symmetry*, 16(5): 2024, 635.

CHAPTER 7

# Postscript to Volume II

Volume II deals with applications. I brought forward the analytical expressions of the power spectrum density (PSD) and cross-PSD responses to seven classes of fractional vibrators driven by fully developed ocean surface waves, fractional Gaussian noise (fGn), generalized fGn (gfGn), fractional Brownian motion (fBm), fractional Ornstein-Uhlenbeck (OU) process, and wind fluctuation speed in Chapters 1–6, respectively. The computation methods with respect to the autocorrelation function (ACF) responses, cross-correlation responses to seven classes of fractional vibration systems driven by ocean surface waves, fGn, gfGn, fBm, OU process, and wind fluctuation speed were addressed in Chapters 1–6. Their analytic expressions are left as exercises.

My intention in Chapters 1–6 is towards possible applications of fractional random vibrations to engineering practice. For example, Chapters 1 and 6 might be meaningful in ships and ocean engineering. As far as engineering applications are concerned, a number of issues are worth paying attention to.

- Test techniques of how to determine an order value of fractional inertia or fractional damping or fractional restoration.
- Sensors and instrumentations for measuring fractional acceleration, fractional velocity, fractional displacement, fractional stress and strain.


DOI: 10.1201/9781003657903-7

- Testing machines of fractional vibrations.
- Testing machines of fractional shocks.

If one feels that this volume may be a help with respect to possible applications of fractional random vibrations, I will be glad.

# Index


A

ACF response, 4, 16, 19, 22, 32, 36, 40, 50, 53, 60, 69, 72, 75, 77, 80, 88, 98, 102, 106, 110, 113, 122, 126, 130, 150, 154, 158, 161

C

class I fractional vibrator, 2, 31, 59, 86, 120, 148
class II fractional vibrator, 6, 35, 64, 90, 124, 152
class III fractional vibrator, 11, 38, 67, 97, 129, 156
class IV fractional vibrator, 14, 42, 69, 101, 132, 158
class V fractional vibrator, 18, 45, 73, 103, 135, 163
class VI fractional vibrator, 21, 48, 76, 107, 138, 165
class VII fractional vibrator, 24, 51, 78, 112, 141, 169
cross-correlation response, 13, 32, 36, 50, 61, 69, 72, 77, 80, 88, 98, 102, 106, 110, 113, 122, 126, 130, 150, 154, 158, 161
cross-PSD response, 4, 8, 16, 19, 22, 24, 32, 36, 40, 43, 47, 50, 53, 60, 64, 68, 71, 74, 77, 80, 88, 92, 98, 102, 106, 109, 113, 122, 126, 130, 133, 136, 139, 142, 150, 154, 157, 161, 164, 167, 170

F

fractional Brownian motion, 85
fractional Gaussian noise, 30
fractional Ornstein-Uhlenbeck process, 119

G

generalized fractional Gaus sian noise, 58

H

Hurst parameter, 31, 59, 86

J

JONSWAP (Joint North Sea Wave Project) spectrum, 27

L

long-range dependence, 54, 82, 85

P

Pierson-Moskowitz spectrum, 1
PSD response, 3, 7, 12, 15, 18, 21, 24, 32, 35, 39, 43, 46, 49, 52, 59, 64, 68, 71, 74, 77, 79, 87, 90, 97, 101, 105, 108, 112, 121, 125, 129, 132, 136, 139, 142, 149, 153, 157, 160, 163, 166, 169

S

short-range dependence, 54, 82, 119, 147, 172

V

von Kármán process, 120, 148
von Kármán spectrum, 147, 148